Cross-linked Polymers as Dielectrics
for Organic Field-effect Transistors

Cross-linked Polymers as Dielectrics for Organic Field-effect Transistors

Vom Promotionsausschuss der
Technischen Universität Hamburg-Harburg
zur Erlangung des akademischen Grades
Doktor-Ingenieur (Dr.-Ing.)
genehmigte Dissertation

von

Zied Fahem

aus Tunis, Tunesien

2013

1. Gutachter: Prof. Dr. Wolfgang Bauhofer
2. Gutachter: Prof. Dr.-Ing. Wolfgang Krautschneider

Tag der mündlichen Prüfung: 22.08.2013

Bibliografische Information der Deutschen Nationalbibliothek

Die Deutsche Nationalbibliothek verzeichnet diese Publikation in der Deutschen Nationalbibliografie; detaillierte bibliografische Daten sind im Internet über http://dnb.d-nb.de abrufbar.

1. Aufl. - Göttingen: Cuvillier, 2013

Zugl.: (TU) Hamburg-Harburg, Univ., Diss., 2013

978-3-95404-561-7

Nonnenstieg 8, 37075 Göttingen
Telefon: 0551-54724-0
Telefax: 0551-54724-21
www.cuvillier.de

1. Auflage, 2013
Gedruckt auf umweltfreundlichem, säurefreiem Papier aus nachhaltiger Forstwirtschaft

978-3-95404-561-7

Contents

1. Introduction

Organic electronics are getting more and more interest from industrial companies and research groups in the last years since they enable many new applications, which could not be realized by inorganic materials [1–7]. Flexible displays [1], large-area sensors [1], light-emitting large surfaces [8], printable radio-frequency identification tags (RFID) for packaging or logistic industry [2] and many other systems which require flexible, large area and low-cost electronic devices are now developed for the near future or even already commercialized. Organic light-emitting-diode (OLED) displays, for example, are now implemented in portable devices and have higher performance than the traditional LCD displays [9]. OLED displays are self illuminating and do not need back lightening, therefore they have higher brightness, contrast and viewing angle in comparison to LCD displays [9]. Many electronic devices producers implemented OLED displays in their high-end smartphones and SLR cameras [10], and recently LG (a Korean company) introduced a 55-inch OLED television [11]. Large-area solar cells based on organic materials have also found their way to commercialization [12].

All of these innovations were only possible after the introduction of organic conductors and semiconductors. Organic (semi)-conductors have the advantage of their low-cost processing technologies (e.g. printing or spray-coating). However, they have lower electrical conductivity, free charge carriers mobility [13] and packaging density than their inorganic counterparts. Therefore they are normally used in low-cost and low-performance applications, except in the case of OLED where they have clear advantages compared with other technologies.

In order to produce fully flexible devices, elementary devices for electronic circuits (e.g. transistors and diodes) need to be made with flexible materials. The performance of these devices needs to be enhanced and their fabrication processes should be optimized to ensure their commercialization as switching elements in OLED displays or in other circuits. These tasks are principally a material issue, that means new materials with higher performance and easier processability are sought after.

Most of the work done in this direction is devoted to synthesize new semiconducting materials with higher mobility and solution-processability. Parallel to these investigations, new dielectric materials with good dielectric properties even in thin films, solution-processability and good interface properties to the semiconductors should be developed. Cross-linked polymers are advantageous for organic field-effect transistors (OFETs) because of their electrical and chemical properties, specially their high volume resistivity and their stability towards solvents, acids and bases. This work shows that the application of cross-linked materials can also reduce or even eliminate the use of the volatile organic compounds (VOC), for which legal limitations are yearly augmented.

1.1 Scope of the thesis

The main goal of this work is to find new dielectric materials for top-gate OFETs. The top-gate configuration was chosen for its benefits for the production of OFETs (see section 2.1 for more details). The dielectric materials should have the following properties:

- Good electrical properties, in particular high volume resistivity
- Solution-processibility to enable low-cost production of OFETs
- Cross-linking so that another solution-processed layer could be deposited on it
- Fast cross-linking reaction to be suitable for high throughput production technologies (e.g. printing)
- Low-temperature processibility to be usable with flexible substrates based on plastics
- Good interface with the semiconductor to have high-performance transistors
- Low or no use of VOC

1.2 Background and guidelines of the thesis

The importance of developing new cross-linked dielectrics for OFETs was noticed after the participation of the institute of optical and electronic materials (OEM) to a € 15 millions scientific and industrial project, "MaDriX". The goals of this project were to: develop printed circuits for RFID, ensure reproducibility, develop techniques for inspection during the production of the circuits and optimize the characterization techniques of printed devices. Different global companies (ELANTAS Beck GmbH, BASF, Evonik Industries AG, PolyIC GmbH & Co. KG and Siemens AG), universities and research institutes with expertise in chemistry, materials for electronics, device development, measurement techniques and in printing were involved in this project. OEM had the task to evaluate dielectric materials developed by ELANTAS Beck GmbH as OFET dielectrics. As other partners were also working on new materials for OFETs, one reference "standard" transistor was needed to compare results. The standard transistor was produced using the same materials and processes by all the partners. The devices were characterized also by standard system using the same measurement parameters, e.g. sweep rates and voltage values. The geometrical structure of the transistors was also standardized. In order to investigate a new dielectric for example, a transistor is fabricated using the new dielectric and the other standard materials in the standard geometry. The fabricated transistor is compared with the standard transistor using the standard characterization system. To be able to directly compare results achieved in different labs, it was necessary to ensure that the different groups measure the same characteristics of the standard transistor in their labs.

In this work, new dielectric material classes, which were not covered by the "MaDriX"-project and which might be fast cross-linked, are investigated. The idea of using a reference transistor was adopted to this study. Uncross-linked polymers were used as comparing reference for the presented materials. The semiconductor used in all the experiments was taken from the same batch to avoid quality fluctuation between different batches. The fabrication process of the devices and the techniques of their characterization are the same for all the devices. Silicon-wafers, functionalized with hexamethyldisilazane, were used as reproducible and reliable substrates.

The characterization of the new cross-linked dielectrics began with the monitoring of their curing reactions, followed by the measurement of their dielectric properties. The study of the interface between the dielectric layer and the semiconductor should be done as part of the selection process. Many classes of materials should be excluded from the selection because of the quality of their interface with the semiconductor, if the low quality of the interface is caused by intrinsic properties of the material. After the characterization of the interface with the semiconductor in a metal-insulator-semiconductor capacitor, transistors were fabricated using the selected dielectric materials.

The characterization of some of the selected and developed dielectric materials were carried by supervised students during their study projects [14, 15] using the previously defined scheme of experiments explained above.

1.3 Structure of the thesis

This work is composed by eight chapters. After the introduction, a theoretical background on OFETs, organic semiconductors and the interfaces between the different layers in OFETs is given. Then, the methods and the experiments used to fabricate and to characterize the devices are shown. The fourth chapter presents the polymers used in this work and their cross-linking mechanisms. In the fifth chapter, dielectric materials based on linear polymers are studied. Then, two different thermally cured polymers were investigated. The seventh chapter deals with photo-curable polymers which were cured very fast and showed good properties in transistors. Conclusions and suggestions for future investigations are presented in the final chapter.

2. Organic field-effect transistors

2.1 Field-effect transistors

A transistor is an elementary electronic device with three electrodes called: gate, source and drain. The current flowing between the source and the drain is controlled by the voltage or current applied to the gate electrode. Therefore, transistors are used in electronic devices as switches or amplifiers. The possibility of miniaturizing solid-state transistors enabled the production of high-density integrated circuits. There are two main classes of transistors: bipolar transistors and field-effect transistors (FETs). The functionality of the latter one will be further explained in this section. There are many possible structures to realize FETs, basically they differ from each other in the way how the controlling electrode (gate electrode) is separated from the semiconductor, for example the gate electrode is separated from the semiconductor by an oxide in metal-oxide-semiconductor FET (MOSFET), by a p-n junction in junction FET and by a Schottky-barrier in metal-semiconductor FET (MESFET). All of these devices are based on crystalline silicon. In 1976, Neudeck *et al.* presented a thin-film transistor (TFT) made with amorphous silicon [16]. The main application of TFTs is as switching elements in active matrix liquid crystal displays. Ten years later, Tsumura *et al.* presented the first polymer-based FET [17]. The presented device consists of a polythiophene electrochemically polymerized between two electrodes. The substrate was a thermally grown SiO_2 on a silicon wafer, the silicon and the oxide play the role of the gate electrode and the insulator. This configuration is widely used to characterize organic semiconductors and to fabricate OFETs.

An OFET can have many possible configurations depending on the order of the deposition of its composing layers. An OFET is called "top-gate" transistor when the dielectric layer is deposited on the semiconductor. Top-contact OFETs have source and drain electrodes deposited on the semiconductor. Fig. 2.1 shows the four possible configurations of OFETs. Electronic and processing aspects should be taken in account during the choice of the appropriate architecture. In bottom-contact configuration, for example, standard processes of photo-lithography could be used to structure the source and drain contacts. The semiconductor is then not exposed to the developing and stripper solution. In the top-gate configuration the dielectric layer and the gate electrode are covering the sensitive semiconductor and in this way protect it from the environmental degrading species [3]. In addition to that, the gate dielectric does not play the role of a substrate for the semiconductor. Two kinds of insulating materials are needed for this configuration: one material with good dielectric properties and good quality interface to the semiconductor and another material acting as substrate and ensuring good orientation of the semiconductor chains for high free charge carrier mobility. In bottom-gate transistors all of these roles should be ensured by a single material, which limits the choice of appropriate materials. In top-gate configuration the two functionalities could be optimized separately.

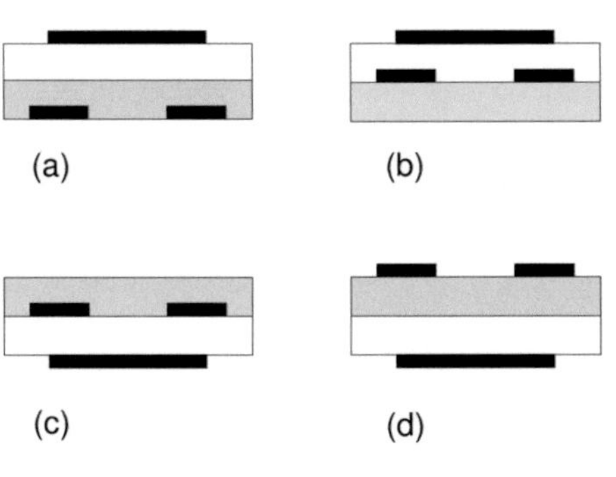

Figure 2.1: Schematics of different OFET configurations (a)top-gate bottom-contacts (TG-BC) (b) TG-TC (c) BG-BC (d) BG-TC. (Metals in black, insulator in white and semiconductor in grey)

The basic principle of function of FETs is that an electrical field can change the density of free charge carriers in the surface of a semiconductor and therefore, also its conductivity. OFETs operate in accumulation regime, it means that the channel induced at the surface of the semiconductor is built with the majority free charge carriers of the semiconductor, i.e. if the semiconductor is p-doped than the channel

is formed by holes. The models and theories adopted for MOSFET, which operates in inversion-regime, still however applicable to OFETs.
The drain current I_d could be described by two equations depending on the operation regime of the FET [18]:

$$- I_{d,lin} = \frac{W}{L} C_i \mu \left(V_{gs} - V_{th} - \frac{V_{ds}}{2} \right) V_{ds} \qquad \text{if } V_{ds} < (V_{gs} - V_{th}) \tag{2.1}$$

$$- I_{d,sat} = \frac{1}{2} \frac{W}{L} C_i \mu \left(V_{gs} - V_{th} \right)^2 \qquad \text{if } V_{ds} > (V_{gs} - V_{th}) \tag{2.2}$$

where $I_{d,lin}$ and $I_{d,sat}$ are the drain current in the linear and saturation regimes respectively, W and L are the channel width and length respectively, μ is the field-effect mobility of the semiconductor, V_{ds} is the drain to source voltage, V_{gs} is the gate to source voltage, V_{th} the threshold voltage and C_i is the capacitance per unit area of the insulator. The transfer characteristic (I_d vs. V_{gs}) is generally used to extract important transistor parameters i.e. field-effect mobility, on/off ratio, V_{th} and sub-threshold swing, as shown in Fig. 2.2. The on/off ratio means the ratio between the currents at the "on" state and the "off" state of the transistor (Fig. 2.2 a). The sub-threshold swing (S) is proportional to the average of the traps at the semiconductor-insulator interface N_{ss}, as shown in the following equation [19]:

$$N_{ss} = \left[\frac{S \ log(e)}{kT/q} - 1 \right] \frac{C_i}{q} \tag{2.3}$$

where e is the Euler's number, k the Boltzmann constant, T the temperature and q the elementary charge. The on/off ratio and the sub-threshold swing could be read from the logarithmic transfer characteristic. The field-effect mobility and the threshold voltage are generally extracted from the slope of the transfer characteristic (I_d vs. V_{gs}) in linear regime or the square rooted transfer characteristic in saturation regime of a FET. However, it was already observed, that the transfer characteristic of amorphous-silicon thin-film-transistors could not be fit with the equations 2.1 and 2.2, because of the dependency of the mobility on the gate voltage [20]. Similar behavior was also observed in OFETs. After Horowitz *et al.* the field-effect mobility could be described by this equation:

$$\mu = \mu_0 \left(V_{go} - V_{th} \right)^{\gamma} \tag{2.4}$$

where μ_0 and γ are empirical numbers [21]. They mentioned that this power-law dependence could be also derivated from the model presented by Vissenberg and Matters, in which the charge transport is governed by the hopping between exponentially distributed localized states [22].

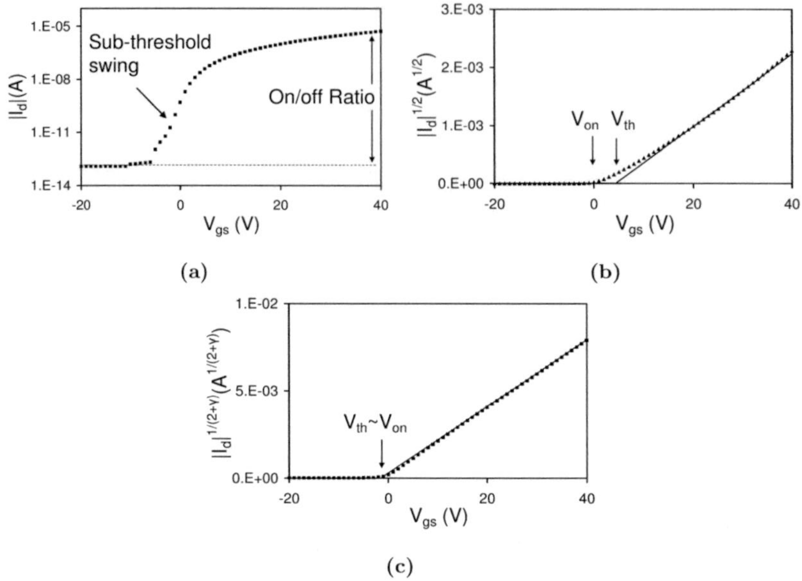

Figure 2.2: (a) logarithmic, (b) square rooted and (c) to $1/(2+\gamma)$ powered transfer characteristics of an OFET.

In their model, the field-effect mobility is proportional to $(V_{gs}-V_{th})$ raised to $\left(\frac{T_c}{T}-1\right)$ where T is the temperature and T_c is a parameter indicating the width of the exponential distribution of the states. γ is then a parameter describing the width of this distribution. The drain current in linear regime is then equal to:

$$I_{d,lin} = \frac{W}{L}C_i\mu_0\left(V_{gs}-V_{th}\right)^{\gamma}\left(V_{gs}-V_{th}-\frac{V_{ds}}{2}\right)V_{ds} \qquad \text{if } V_{ds} < \left(V_{gs}-V_{th}\right)K \quad (2.5)$$

and in saturation regime to:

$$I_{d,sat} = \frac{1}{2+\gamma}\frac{W}{L}C_i\mu_0\left(V_{gs}-V_{th}\right)^{2+\gamma} \qquad \text{if } V_{ds} > \left(V_{gs}-V_{th}\right)K \quad (2.6)$$

where $K = \left(1-\sqrt{\frac{\gamma}{2+\gamma}}\right)$. It is noticeable that the value of V_{ds} separating the linear and saturation regimes is changed by the factor $K = \left(1-\sqrt{\frac{\gamma}{2+\gamma}}\right)$, which means that the transistor is already in saturation regime at drain to source voltages lower than $(V_{gs}-V_{th})$. The drain current should be then raised to $\frac{1}{2+\gamma}$ in order to extract the field effect mobility and the threshold voltage (Fig. 2.2 c). In this curve, the extracted threshold-voltage is near the onset-voltage as in the case of the ideal FET.

2.2 Organic semiconductors

2.2.1 Origin of conductivity in conjugated polymers

Organic materials were introduced first in electronic devices as insulators or as packaging materials, because of their good insulation, processability and mechanical properties. In 1977, it was discovered that an insulating polymer, trans-polyacetylene, with low conductivity ($10^{-3}\,S\,cm^{-1}$) becomes highly conducting after exposure to an oxidizing agent [23]. After this breakthrough in the area of conducting organic materials, other polymers with the same properties were discovered, for example poly(p-phenylene), polypyrrole, polythiophene, polyfuran and their derivatives. Many of these materials are used as semiconductors in the undoped state, for example, the first presented OFET was based on undoped polythiophene [17].

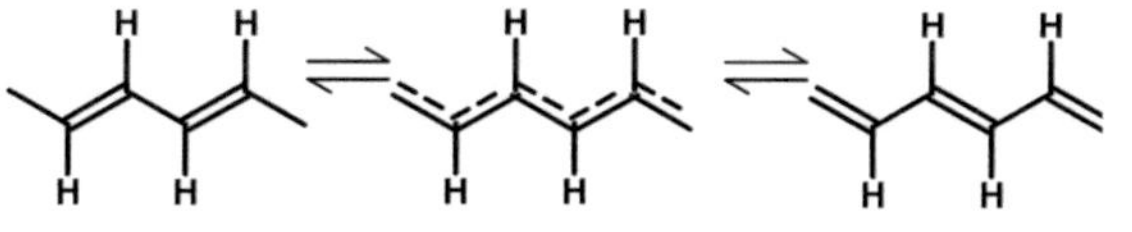

Figure 2.3: The two degenerate A and B phases of *trans*-polyacetylene.

The (semi-)conducting organic materials are based on conjugated systems, that means they contain alternating single and multiple bonds [24]. The "backbone" of the polymer is built by in-plane σ-bonds. Every carbon atom of the conjugated chain still has one unpaired electron (π electron). The orbitals of the π electrons of adjacent carbon atoms overlap and build the π-bond, which leads to an electron delocalization along the polymer. The full bonding orbital (π-bond) and the empty anti-bonding orbital (π^*-bond) correspond to the highest occupied molecular orbital (HOMO) and to the lowest unoccupied molecular orbital (LUMO) respectively. The existence of such bands is one requirement to have electrical conductivity. The other requirement is to have free charge carriers in the material. In inorganic semiconductors, free charge carriers are generated by taking electrons from the valence band (generation of holes) or by donating electrons to the conduction band. The molecular structure of an organic polymer has high interaction with the charge added to it [25]. Because of this interaction, quasi-particles are expected to be responsible for the electric conductivity in organic polymers.

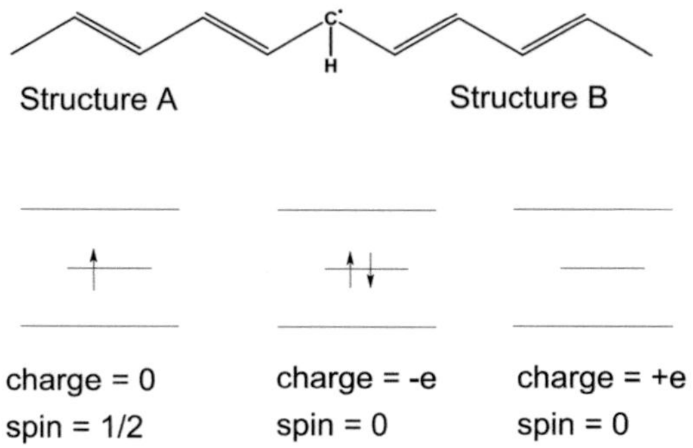

Figure 2.4: Structure of a soliton in a *trans*-polyacetylene and its gap state occupancies in dependence of its charge [24].

The conducting polymers could be divided into two different classes: polymers with degenerate ground state and polymers with non-degenerate ground state. trans-Polyacetylene is a polymer with a degenerate ground state, i.e. it has two geometric structures A and B having the same total energy (see Fig. 2.3) [24]. These two structures, in which the double bonds are shorter than the single bonds, are more stable than the structure with equidistant carbon atoms because of the Peierls instability [24]. When a defect in the bond alternation occurs, a carbon atom would have an unpaired electron and it will separate two branches of the polymer having the two different structures A and B (see Fig. 2.4). These defects are called solitons, the unpaired electron would have a new energy level at the mid gap [24, 26]. The neutral soliton has a spin of 1/2, since the carbon atom is occupied with only one electron. The soliton has no spin when it is positively or negatively charged (see Fig. 2.4) [24].
All the other conjugated systems have non-degenerate ground states [26]. In these polymers, other quasi-particles are responsible for the electrical conductivity: polarons and bipolarons. Polarons could be presented as a neutral soliton associated with a charged soliton, and the bipolarons as the association of two charged solitons [24]. Polythiophenes, for example, are conjugated polymers with a two non-degenerate ground states: an aromatic and a quinoid structures (Fig. 2.5 a,b) [26].

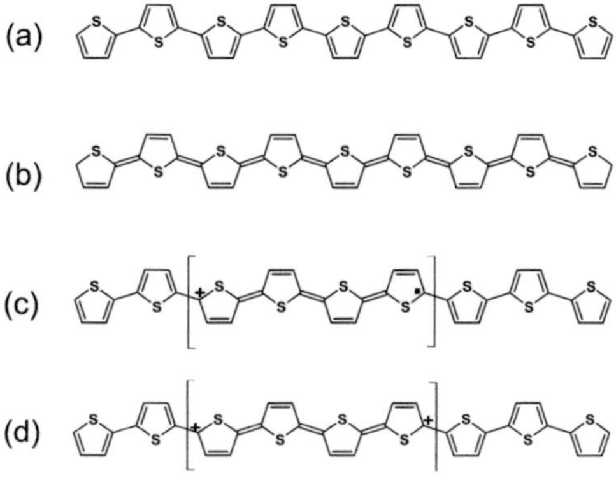

Figure 2.5: Aromatic (a) and quinoid-like (b) geometric structures of polythiophenes. Illustration of the formation of a positively charged polaron (c) and bipolaron (d).

The aromatic state has the lowest energy [26]. Taking one electron from the polythiophene leaves one cation and a unpaired electron associated with a change in the local structure from the aromatic to the quinoid form (Fig. 2.5 c). Further oxidation of the polymer leads to the formation of two cations in the chain associated with a structure distortion, this quasi-particle is called bipolaron (Fig. 2.5 d). Each of these cations separates two domains with two different ground states (aromatic and quinoid), and therefore could be seen as positively charged solitons. Polarons and Bipolarons could be positively or negatively charged and have a spin of 1/2 and 0 respectively (Fig. 2.6) [24].
The terminology of inorganic semiconductor physics is widely used to describe systems based on organic polymers. For example HOMO and LUMO could be called valence and conduction bands respectively. The oxidation or reduction of the polymers (taking or adding electrons to the system) designated as doping process. It is however important to remember, as already shown, that the nature of the free charge carrier is different. In inorganic semiconductors the dopant atom is incorporated in the crystal structure and exchange electrons with the bands of the semiconductor. After the doping process one hole is left in the valence band or one electron in the conduction band. However, when one electron is taken from an organic semiconductor, the ionization energy is reduced [25]. The unpaired electron has an energy level

localized in the forbidden region, and no free charge carrier is added to the HOMO or LUMO of the polymer [25].

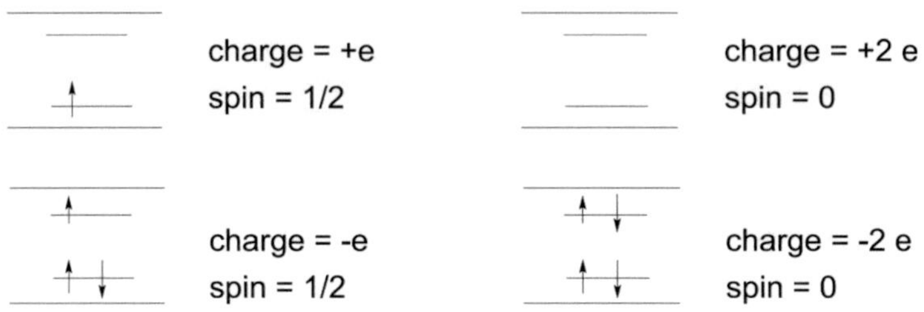

(a) (b)

Figure 2.6: Energy level and spin of positive and negative (a) polarons and (b) bipolarons [24].

2.2.2 Solution-processed organic semiconductors

Organic semiconductors with high mobilities are mostly based on small molecules and oligomers. Films based on pantacene or rubrene achieved field effect mobilities similar to amorphous silicon. These films are generally produced using thermal evaporation in vacuum. Therefore, they are not suitable for applications, in which such expansive processes could not be used. The production cost of electronic circuits could be reduced by introducing materials which are processable by large-area growth techniques like printing. To achieve this, solution-processed organic semiconductors are needed. One strategy to have soluble organic semiconductors is to build substitutions with alkyl-groups. Some of commercially available soluble semiconductors are shown in Fig. 2.7. Poly(3-hexylthiophene) (P3HT) is the most studied soluble organic semiconductor. The side-groups (the hexyl chains) of P3HT enable its dissolution in many non-polar solvents like chloroform, THF, xylene and toluene.

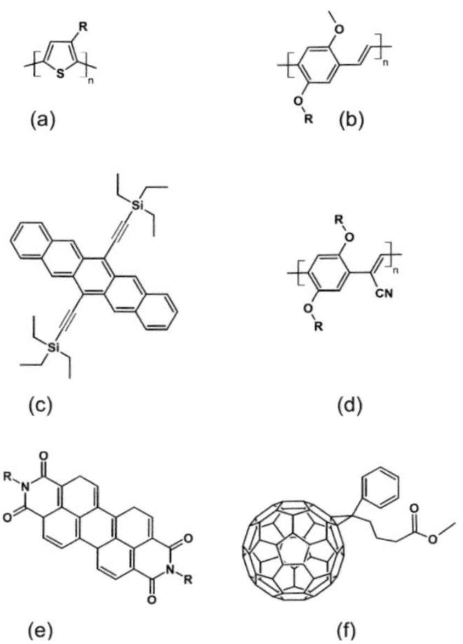

Figure 2.7: Structures of some commercially available soluble organic semiconductors: (a) poly(3-alkylthiophene) ($R = C_6H_{13}$) (b) Poly(2-methoxy-5-alkoxy 1,4-phenylenevinylene) (c) 6,13-Bis((triethylsilyl)ethynyl)pentacene (d) Poly(2,5-di (alkoxy) cyanoterephthalylidene) (e) N,N'-alkyl-perylene dicarboximide (f) Phenyl C_{61} butyric acid methyl ester (PCBM).

2.3 Solution-processed dielectric materials

Because of its availability from conventional silicon-based electronics, silicon dioxide is widely used as dielectric in OFETs to test organic semiconductors, optimize the processing and model the transport in organic semiconductor and the effect of the interfaces in OFETs. In 1990, Peng *et al.* showed that organic insulators can improve the performance of OFETs made with alpha-sexithienyl [27]. They reported that cyanoethylpullulan improved the field-effect mobility over two order of magnitude in comparison with SiO_2. In the 1990s, other reports on organic insulators for OFETs are published [28–31]. The motivation of those works was that organic insulators enable low-cost fabrication of flexible and large-area electronics. OFETs

based on polyethylene terephthalate (PET) [28], poly(4-vinylphenol) (PVP) [29], polyimide [30,31] and poly(methyl methacrylate) (PMMA) [31] were studied.
Using an uncross-linked polymer limits the choice of the material of the layer deposited over it. It is important then to find an orthogonal solvent for each layer. In 2000, Gelinck *et al.* reported a cross-linkable dielectric which could be used in bottom-gate OFETs based on solution-processed semiconductor [32]. In this configuration (back-gate), high chemical stability of the dielectric layer is needed. They used a commercially available photoresist and fabricated bottom-gate transistors based on pentacene, polythienylenevinylene, and regioregular P3HT.
Klauk *et al.* presented another cross-linkable system based on poly(melamine-co-formaldehyde) (PMF) and PVP. After curing at 200 °C, the films have good insulating and chemical properties. The reported pentacene-based OFETs have high field-effect mobilities of up to 3 $cm^2\ V^{-1}\ s^{-1}$. P3HT-based OFETs having PVP-PMF blends as dielectric showed also high field-effect mobilities (0.1 $cm^2\ V^{-1}\ s^{-1}$) [33]. Insulators based on PMF-PVP blends are widely used because of their high performance in OFETs [34–36]. Choi *et al.* showed that the cross-linking reaction between PMF and PVP could occur at temperatures below 100 °C. They cured PVP-PMF at 70 °C for 60 min under vacuum (10^{-2} Torr) to fabricate pentacene-based OFETs [37]. PVP could be also cross-linked by an esterification reaction with bifunctional anhydrides, acyl chlorides, and carboxylic acids at 60 °C - 100 °C for 2 h [38]. Other polymer systems which are cured by esterification reactions are reported. Xu *et al.* showed that the cross-linking of a polyacrylate copolymer including epoxy and carboxylic acid groups improves the dielectric properties and the performance of the fabricated OFETs. They cured the films at 170, 200 and 230 °C for 30 min. The OFETs made with uncured insulator have high hysteresis in the transfer characteristic. The hysteresis is reduced by curing the insulator and becomes negligible after curing at 230 °C [39]. They showed also that PMF mixed with cyanoethylated pullulan and cured at 200 °C for 1 h, have high-k value (10-15) and could be used to fabricate bottom-gate pentacene-based OFETs [40,41]. Divinyltetramethyldisiloxane-bis-benzocyclobutene (BCB) is another material which cures thermally at temperatures higher than 200 °C and which has high performance as insulator in OFETs [42].
Most of the reported thermally cured insulators need high temperature processes which are in many applications not desired e.g. for flexible electronics on plastic substrates. Photo-cured insulators, however, could be cured at lower temperatures

without distorting the substrates. Photo-cured insulators enable also the direct patterning of the gate dielectric. Lee *et al.* cross-linked PVP with a trifunctional epoxy by using triphenylsolfonium triflate as a photo-acid-generator (PAG) (Benzoyl peroxide was also added to enhance the network formation) [43]. They cured the films under 600 W UV light for 10 min and then at 100 °C for 30 min. They showed that the insulator could be patterned using photo-lithography because of its stability to solvents and bases used in the photo-lithography process. Bottom-gate pentacene-based OFETs made with this photo-curable insulator showed lower leakage currents and higher field-effect mobilities and on/off ratios than OFETs made with bare PVP. Other PVP-based formulations containing different PAGs and/or cross-linking agents (CLAs) were reported to be used as insulating layers in OFETs. For example Yang *et al.* used PMF as CLA and a phenyliodonium hexafluoroantimonate salt as PAG and cured the films for 1 min under 356 nm UV-irradiation (9.6 mW cm^{-2}) and then for 30 min at 120 °C [44]. 1,2,4,5-Tetraacetoxymethylbenzene was used also as CLA with 2,4,-bis(trichloromethyl)-6-aryl-1,3,5-triazine as PAG to get photopatterned gate dielectric for OFETs [45]. Other photo-curing mechanism, which do not involve the highly corrosive acids, were also used to fabricate polymeric insulators. Poly(vinyl cinnmate) (PVCi) cures, for example, by (2 + 2)-cycloaddition under UV-irradiation to a cross-linked matrix [46]. Pentacene-based OFETs fabricated with PVCi show no hysteresis, high field-effect mobilities and the cross-linked PVCi have low leakage currents [47, 48]. The PVCi films needed 20 min of exposure to UV-lamps to be cured. Unlike PAG based formulation, the curing of PVCi does not need any thermal treatment, i.e. the reaction of curing of PVCi stops when the UV-irradiation is shut down. Insulators based on acrylic-based photoresist [49], photosensitive polyimide [50, 51] and acrylate-functionalized PMF [52] were also shown to be suitable for high performance bottom-gate OFETs after they have been cured for some minutes under UV-irradiation.

Solution-processed inorganic dielectrics could be prepared by using sol-gel processes. Cahyadi *et al.* used sol-gel prepared SiO_2 as a capping layer on thermally grown SiO_2 gate insulator to improve the field-effect mobilities and the on/off ratios of pentacene based OFETs. Tetraethylorthosilicate was used as precursor, it was mixed with solvents and a catalyst and thermally treated at 100 °C in 10^{-7} mbar [53]. Many other oxides were used because of their high-κ values. Hafnium oxide was prepared from a solution of hafnium tetraethoxide treated for 1 h at 450 °C. The resulted films have a dielectric constant equal to 11 and were successfully imple-

mented in a pentacene-based OFET [54]. Ramajothi *et al.* showed that it is possible to have room-temperature solution-processed inorganic dielectric [55]. They used titanium(IV) propoxide as precursopr and HCL as catalyst. The P3HT-based OFETs prepared with the sol-gel processed TiO_2, have on/off ratios between 10 and 100. The authors claim that the low on-off ratios could be caused by the high surface roughness of the insulating layers [55].
Inorganic polymers based on silsesquioxanes were also reported to be suitable for high performance OFETs [56]. The spin-coated films were heated for 12 h at 135 °C. Jeong *et al.* used a precursor solution based on methytriethoxysilane and tetraethylorthosilicate mixed with water and an acid catalyst to synthesize an insulating film based on polysiloxane [57]. They showed that curing the film at 190 °C is needed to get hysteresis-free bottom-gate transistors. At lower curing temperature, a higher concentration of silanol was measured by FTIR, causing a hysteresis in the transfer characteristic of the transistor.

2.4 Interfaces

The performance of an OFET is highly dependent on the interfaces between the semiconductor and the other layers. In top-gate architecture, there is three important interfaces which should be taken in account: the semiconductor/electrodes, semiconductor/substrate and the semiconductor/dielectric interfaces. In this section, the importance and the methods of their engineering are presented.

2.4.1 The semiconductor/electrodes interface

The injection of free charge carriers from and into the organic semiconductor is limited by the mismatch of the work function between the electrodes and the channel. The use of self-assembled monolayers (SAM) deposited on the electrodes can reduce the contact resistance to the active layer [58]. Because of the strong affinity of sulfur to noble metals, generally SAM are grown on gold electrodes by using thiol-functionalized molecules.

2.4.2 The semiconductor/substrate interface

It is important to notice, that the effects related with this interface are reported together with the effects related to the semiconductor/dielectric interface, since the majority of the published works deals with bottom-gate transistors. The semiconductor/substrate interface is important in controlling the intrinsic properties of the grown semiconductor films since the morphology of the films depends on the surface tension of the substrate [59, 60]. Different approaches were presented to change the chemical nature of the surfaces of the substrate used for the organic semiconductor growth. Alkoxysilanes, trichlorosilanes [61–63], phosphonic acids [64] and hexamethyldisilazane (HMDS) [65] were used to functionalize surfaces based on inorganic oxides like silicon dioxide, titania and alumina. The silanes and the phosphonic acids with alkyl-terminal group are generally used, because the alkyl termination reduces the surface free energy of the substrates, which is beneficial for increasing the field-effect mobilities of the deposited semiconductors.

2.4.3 The semiconductor/dielectric interface

The properties of this interface have high impact on the electrical characteristics of OFETs, since the majority of the source-drain current is flowing in the first several layers of the semiconductor close to the insulator. The quality of the semiconductor/dielectric interface has an influence on the field-effect mobility of the free charge carriers, the threshold / flat-band voltage of the devices and their sub-threshold characteristics.

The chemical and electrical properties of the dielectric material which is in contact with the semiconductor have influence on the free charge carrier mobility. It was reported that polar groups (i.e. in high-k dielectrics) at the interface between the semiconductor and the dielectric material decrease the field effect mobility [66]. The presence of OH-groups at the interface can discriminate the n-type conduction in the active layer as they act as electron traps [67, 68]. Lee *et al.* showed that OH-groups at the interface can also increase the hole mobility [69].

In the other hand, trapped and fixed charges available at the semiconductor/dielectric interface in a field-effect transistor can change the threshold-voltage and the subthreshold behavior [18]. In bottom-gate transistors, it is possible to control the threshold voltage by growing self-assembled layers containing functional groups acting as dipoles [70, 71]. This technique is not available for top-gate transistors. There are only two ways to control the characteristics of the interface between the semiconductor and the dielectric layer: change the insulating material or separate the semiconductor and the dielectric with another material having better interfacial properties.

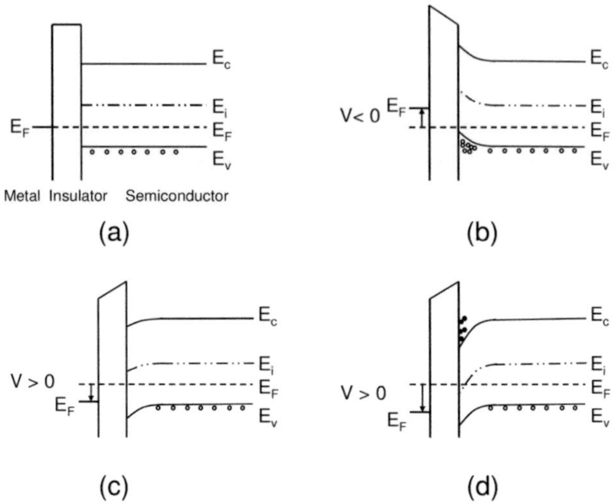

Figure 2.8: Band structure of a MIS-capacitor in (a) flat-band conditions, (b) accumulation, (c) depletion and (d) inversion.

2.4.4 CV-characteristics

In the following section, an appropriate method to characterize the semiconductor/dielectric interface is presented. The characterization of metal-insulator-semiconductor (MIS)-structures is a powerful tool to have informations about the interface between the semiconductor and the dielectric material. Many data about the insulator-semiconductor interface could be obtained from the measurement of capacitance of MIS-structures in dependence of the applied dc bias-voltage. Fig. 2.8 (a) shows the band diagram of an insulator sandwiched between a metal and a semiconductor in equilibrium. When the applied voltage is zero, and assuming that the work function of the metal and semiconductor have the same value, the valence band and the conduction band of the semiconductor are flat. For this reason this voltage value is called flat-band voltage (V_{fb}). When a negative voltage is applied (Fig. 2.8 (b)), the valence band near the interface is bent so that the difference between the valence band level and the Fermi level near the interface is reduced, and therefore, the density of holes is increased. The MIS-capacitor is then in accumulation. The total capacitance is equal to the insulator capacitance. When a positive voltage is applied (Fig. 2.8 (b)), the valence band near to the interface is bend so that the difference between the valence band level and the Fermi level near the interface is increased, and therefore, the density of holes is decreased (see Fig. 2.8 (c)). The area near the interface is then depleted from free charge carrier. The depleted area is thicker the higher the applied voltage. Since the total capacitance is equal to the serial connection of the capacity of the insulator and the capacity of the depleted area, the capacity of the device is then voltage dependent and lower than the capacity of the insulator. At higher positive applied voltage, the Fermi level becomes nearer to the conduction band than to the valence band. In this case, the electron density near the interface is increased (see Fig. 2.8 (d)). The MIS is then inverted, and the capacitance of the MIS is equal to the insulator capacitance. The inversion can only be measured when the generation of the electron is faster than the ac signal used to measure the capacitance. At high frequency, no inversion is observed in the CV measurements. In the case of organic semiconductor, no inversion could be measured even at low frequencies of some Hz [72, 73]. That means that the thermal generation rate of the free charge carrier in organic semiconductor are much slower.

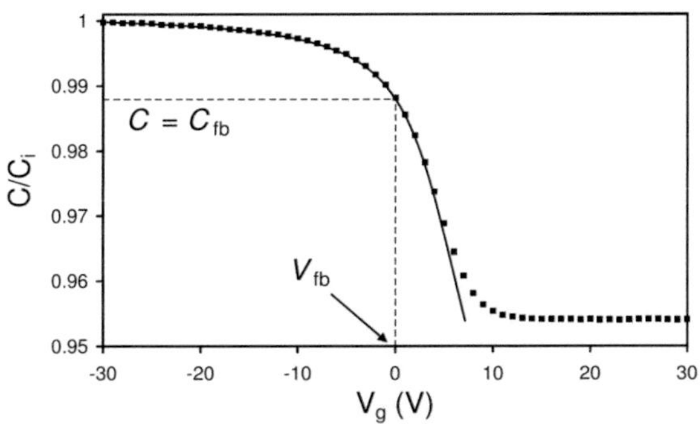

Figure 2.9: Typical CV-characteristics of an organic MIS-capacitor.

Fig. 2.9 shows a typical CV-characteristic of a MIS capacitor made with an organic semiconductor. As already discussed, the capacitance of the MIS-structure is dependent on the applied voltage. The CV curve could be described by theoretical equations. However there is no explicit equation giving the capacitance as a function of the applied voltage. Instead of that, the capacitance and the voltage could be described as functions of the surface potential (ϕ_s) of the semiconductor. In the following equations, all the capacitances and the charges are divided by the area. The capacitance of the MIS capacitor is equal to the serial connection of the insulator capacitance (C_i) and the capacitance of the depletion region C_D:

$$\frac{1}{C} = \frac{1}{C_D} + \frac{1}{C_i} \tag{2.7}$$

The capacitance of the depletion region is dependent on the surface potential (ϕ_s) and the surface charge (Q_s) [74]:

$$C_D = \frac{dQ_s}{d\phi_s} \tag{2.8}$$

where

$$Q_s = \pm\sqrt{2\epsilon_s kTN_a}\left[e^{-q\phi_s/kT} + \frac{q\phi_s}{kT} - 1\right]^{\frac{1}{2}} \tag{2.9}$$

ϵ_s is the dielectric constant of the semiconductor, k the Boltzmann constant, N_a the doping concentration, q the elementary charge and T is the temperature. In the equation 2.9, only the majority free charge carrier are taken in account. The gate voltage is equal [74]:

$$V_g = \frac{|Q_s|}{C_i} + \phi_s \tag{2.10}$$

To plot the theoretical curve it is only necessary to calculate for every value of Φ_s the corresponding values of applied voltage and total capacitance and plot the point with these coordinates. It could be demonstrated that the slope of the plot of $\frac{1}{C^2}$ versus the applied voltage is proportional to the doping concentration [19]. The doping concentration could be than calculated by this equation [19]:

$$N_a = \frac{2}{q\epsilon_s}\left[\frac{d\,(1/C^2)}{dV_g}\right]^{-1} \tag{2.11}$$

The gate voltage dependence on ϕ_s is changed when the device has charges in the dielectric layer or its interface with the semiconductor. There are four types of charges which can exist in MIS-structures:

- Fixed charges which do not depend on the applied voltage. Fixed charges in the dielectric layer cause a parallel shift of the CV-curve. The charge density of the fixed charges is proportional to the V_{fb}-shift (ΔV_{fb}) [19]:

$$Q_f = -\Delta V_{fb}\, C_i \tag{2.12}$$

- Mobile ionic charges originating from contaminations. Ion migration can follow the field and increases the polarization in the dielectric layer. The mobile charge can cause a clockwise hysteresis in the CV-curve of a p-type MIS capacitor based on polymers. When a negative voltage is applied, anions are accumulated near the interface between insulator and semiconductor. When the voltage is swept to 0 V, the anions stay at the interface and hold the positive charge carriers accumulated in the semiconductor near to the interface to the dielectric layer.
- Oxide trapped charges: These charges could be injected from the semiconductor or the gate electrode. They can also cause a shift in the CV curves.
- Interface trapped charges: These traps are discrete level in the forbidden gap of the semiconductor at the interface with the dielectric layer. There are two kind of traps: The donor-like traps which become positively charged after donating an electron and the acceptor-like traps which become negatively charged after accepting an electron. Their charge depends on the potential at the interface. Traps could have different response times to a change of potential depending on their nature. The fact that a trap has a response time (traping and detrapping rate) means that the behavior of interface traps is strongly dependent on the frequency of the excitation signal. At low frequencies, the traps could follow the signal and have a contribution in the total capacitance of the MIS capacitor. At high frequency, traps can neither trap charge nor release it. There are two possibilities to use CV-measurements to quantify the density of interface traps. The first one is to compare the CV-curves at low and high frequency. The difference between the two curves is caused by interface traps [19]. The second method , the so-called Terman method, compares the high frequency curve with the theoretical one [75]. The CV curve is stretched due to the interface traps, since their charge changes during the sweep of the applied dc bias-voltage. To determine the density of the interface traps, the theoretical surface potential and gate voltage corresponding to every measured capacitance value should be calculated. The density of states is described by [19]:

$$D_{it} = \frac{C_i}{q} \frac{d(\Delta V_G)}{d\Phi_s} \tag{2.13}$$

where ΔV_G is the difference between the measured value of the gate voltage and the theoretically ideal one.

3. Methods and experiments

General methods and experiments which are used to investigate all the dielectric materials presented in this work, are explained in this chapter. Specific methods and experiments related to each dielectric material are explained in the corresponding sections.

3.1 Methods

3.1.1 Monitoring of the curing processes

The curing of a formulation occurs through chemical reactions between their functional groups. To monitor the curing process, any physio-chemical property of the formulation, which change during the reaction, can be used. The reaction could be directly investigated by measuring the decreasing of a specific property of the functional group or the increasing of a specific property of the products. For this purpose spectroscopic methods like Fourier-transform infrared spectroscopy (FT-IR), UV-Vis or Raman spectroscopy could be used [76]. The curing process could be also monitored by measuring for example the dielectric properties of the formulation [77] or the heat generated during the reaction [78].

In this work, the polymerizations are studied by FT-IR. If the spectra do not deliver the needed data, the curing process is investigated by measuring the dielectric constant, loss and the volume resistivity. FT-IR spectra of the solid films spin-coated onto double-side polished silicon wafers were recorded in transmission mode using a Bruker Tensor 37 FT-IR Spectrometer before and after the curing. Metal-insulator-metal (MIM) were made with the investigated polymers as showed in the following subsection, and their electrical properties were measured using HP 4284A Precision LCR meter at 1 kHz and an Agilent 4156C Precision Semiconductor Parameter Analyzer during the curing process. The films, used in the characterization of the curing process, should have the same thickness as in transistors, since the speed of the curing could be dependent on the film thickness.

In the case of monitoring the disappearance of band specific to a functional group involved in the reaction, the degree of cure can be expressed as shown in this equation:

$$Degree\,of\,cure = \frac{A_0 - A_t}{A_0} \tag{3.1}$$

Where A_0 is the ratio of the integrated area of the peak specific to the functional group involved in the reaction and the area of a reference peak before irradiation; A_t is the ratio of the areas of the same peaks at time t. A reference peak is a peak specific to a functional group which does not change during the reaction.

3.1.2 MIM- and MIS-capacitors characterization

The electrical characterization of MIM-structures is performed in this work in order to determine the dielectric constant and the volume resistivity of the films. The dielectric constant was determined by measuring the capacitance of a MIM-structure by using a HP 4284A LCR-meter. The dielectric constant were calculated by using the following equation:

$$\epsilon_r = \frac{C\,d}{\epsilon_0\,A} \tag{3.2}$$

where C is the measured capacitance, d the thickness of dielectric layer, ϵ_0 the permittivity of vaccuum and A the area of the MIM-capacitor. The volume resistivity is determined by measuring the current-voltage (IV) characteristic of the capacitor with Agilent semiconductor parameter analyzer 4156C. The current flowing through the capacitance consists of two parts: a charging current which decreases with time and the leakage current which is limited by the resistivity of the insulator. Therefore for every voltage value it is important to wait till the current flowing through the insulator stabilizes before measuring at another voltage. The volume resistivity is calculated using the following equation:

$$\rho = \frac{V\,A}{I\,d} \tag{3.3}$$

where V is the applied voltage, A the area of the MIM-capacitor,I the measured current and d the thickness of dielectric layer.

CV-characteristics of MIS-capacitors were measured with a HP 4284A LCR-meter at 1 kHz. The measurement were done in air, in vaccuum (10^{-2} mbar) or in nitrogen atmosphere.

3.1.3 Characterization of transistors

After the CV-characterization of the MIS capacitors and the optimization of the V_{fb}, corresponding OFETs were fabricated and characterized using Agilent semiconductor parameter analyzer 4156C in nitrogen atmosphere. The transistor were annealed before the measurement for 30 min at 90 °C in nitrogen atmosphere.

3.2 Preparation of the devices

The dielectric materials were prepared for spin-coating by dissolving them in organic solvents when necessary. The solutions were filtered with 0.45 μm PTFE-syringe filters. For MIM structures (see Fig. 3.1 (a)), Cr and Au were evaporated with electron-beam evaporator on glass substrates. The dielectric materials were spin-coated on the metal coated substrates. Polymers were annealed at 90 °C, and the monomer formulations were cured by the appropriate energy source. Then gold was evaporated through a shadow mask to get the upper electrode.

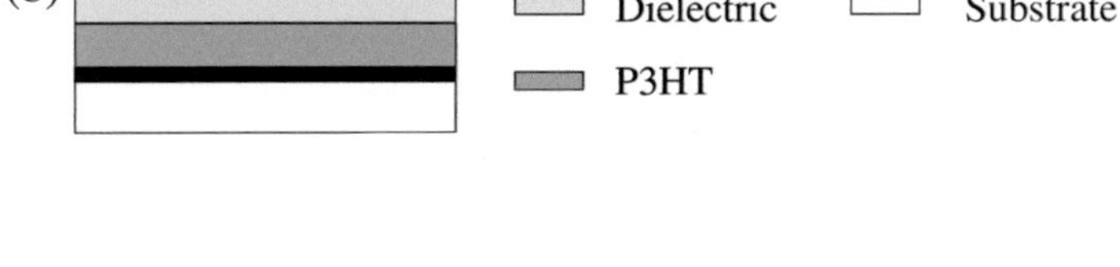

Figure 3.1: Schematic cross-section of (a)the metal-insulator-metal structure, (b) the metal-insulator-semiconductor capacitor and of (c) the top-gate transistor used in this work.

The MIS-structures (see Fig. 3.1 (b)) were fabricated in the same way like the MIM-structures with an additional P3HT-layer. P3HT (supplied by BASF) was dissolved in toluene (1.5 wt.%), filtered like the dielectric materials and then spin-coated on the substrates to have about 50 nm thick films. The P3HT layers were dried on a hot plate for 1 min at 90 °C. The dielectric material was then spin-coated on the semiconductor layer.

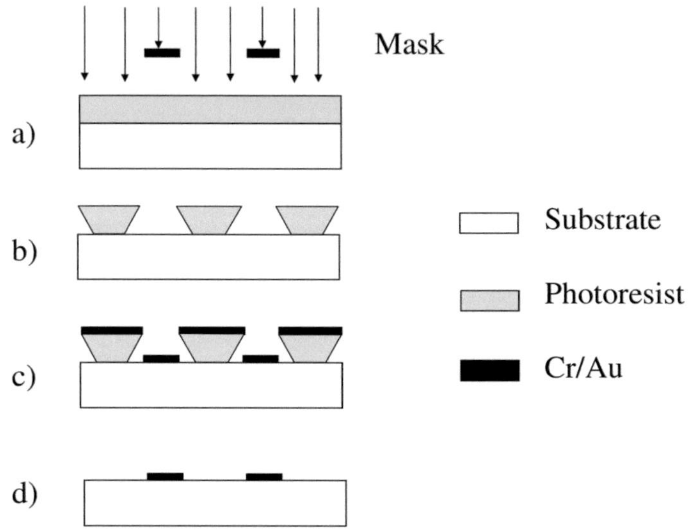

Figure 3.2: Lift-off process steps: a)and b) spin-coating of the photo-resist and its patterning, c) deposition of a Cr/Au layer and d) remove of the photo-resist with acetone.

For the transistors (see Fig. 3.1 (c)), silicon wafers with 300 nm silicon oxide were used as substrates. First they were immersed for 30 min in a 25 % aqueous solution of ammonia, then washed with ultrapure water, dried with nitrogen gun and heated to 100 °C for one minute. The wafers were then immersed for 6 h in 1 mM solution of HMDS in acetone. The functionalized wafers were washed in aceton in ultrasonic conditions to take away the non reacted HMDS. Then, source and drain interdigitated electrodes were prepared using a lithography/Cr-Au deposition/lift-off process. For this purpose, a negative tone photo-resist (ma-N 1400 purchased from micro resist technology) was used as sacrificial layer. It was spin-coated on the wafers for 30 s at 3000 rpm. The coated wafers were heated on a hotplate for

90 s at 100 °C, and then exposed to an i-line UV-irradiation (450 mJ cm^{-2}) using a photo-mask. The wafers were then developped for 70 s in ma-D 533/S, rinsed with dionized water and dried. After the deposition of Chrom and Gold, the sacrificial layer was removed with acetone in ultrasonic conditions. In this way structured metal electrodes are left after this process. The steps of the lift-off process are shown in Fig. 3.2. The thickness of the source and drain electrodes was 50 nm. The channel length and width were 20 and 6260 μm, respectively. The source/drain electrodes are shown in Fig. 3.3.
Then P3HT (0.7 wt.% in xylene, spincoating parameters: 2000 rpm and 15 s leading to a thickness of about 20 nm) was deposited as active layer. Xylene was used because after prelimenary tests, P3HT dissolved in xylene had higher field-effect mobilities than samples prepared using toluene and other solvents with lower boiling temperatures. The dielectric layers were deposited in the same way done for MIM-structures. Gold was evaporated through a shadow mask to get the gate electrode.

Figure 3.3: Optical microscopy picture of an interdigitated source/drain electrodes. The length of the channel and the width of the digits are 20 μm and 6260 μm respectively. The length of each digit is 1 mm.

4. Selected cross-linked dielectrics: mechanisms and raw materials

The cross-linked materials, presented in section 2.3, need long-time curing and/or temperatures higher than 150 °C. A deeper view at the possible dielectric materials and their mechanisms of curing is needed to find new materials, which could be cured faster and at lower temperatures. Two cases are discussed in the following sections: the case of thermally cured polymers and the case of photo-curable polymers. Two approaches are presented here for thermally curable polymers: the first one is based on the optimization of a linear polymer investigated in the chapter 5 and the second one is based on polysiloxane. The approach of using photo-polymerizable monomers based on acrylates or on thiol-ene as gate dielectric for top-gate OFETs is, however, novel and therefore explained in the following sections in more details. The curing chemistry and the possible raw materials are presented for each approach.

4.1 Thermally cured polymers

4.1.1 Cross-linked reactive polymers

Most of the cross-linked polymers used in OFETs are based on at least one polymer with reactive side-groups (see section 2.3). Phenolic resins were reported to cross-link with PMF and with epoxides in the presence of an acid. The time of curing of these formulations was longer than 1 hour. There are many other possibilities to cross-link polymer with reactive side-groups. The first experiments done with linear polymers showed that high performance OFETs could be fabricated with poly(2-vinylpyridine). The latter polymer is a Lewis base and can be cross-linked with multi-functional epoxides. The mixture could be cured fast at temperatures between 100 and 150 °C.

4.1.2 Polysiloxanes

Polysiloxanes, as shown in section 2.3, are suitable for solution-processed OFETs. Their curing needs high temperatures exceeding 150 °C or/and long times, generally hours. The use of a prepolymer containg methoxy substitutions and an adequate catalyst could reduce the curing time and temperature. Silres MSE 100 (Wacker AG), which is a commercially available methoxy functionalized polymethoxysiloxane, could be (according to the producer) cured at room temperature for 15 min with acids or bases as catalysts. The producer recommend the use of an organo-titanate lewis acid.

The mechanism of the curing of this polymer is based on two steps: hydrolysis (Fig. 4.1 a) and condensation (Fig. 4.1 b-c). In the hydrolysis step, the alkoxysilane is transformed in presence of water and an acid or a base into a silanol. The condensation reaction occurs between two silanols producing one siloxane and water or between one silanol and an alkoxysilane producing siloxane and an alcohol.

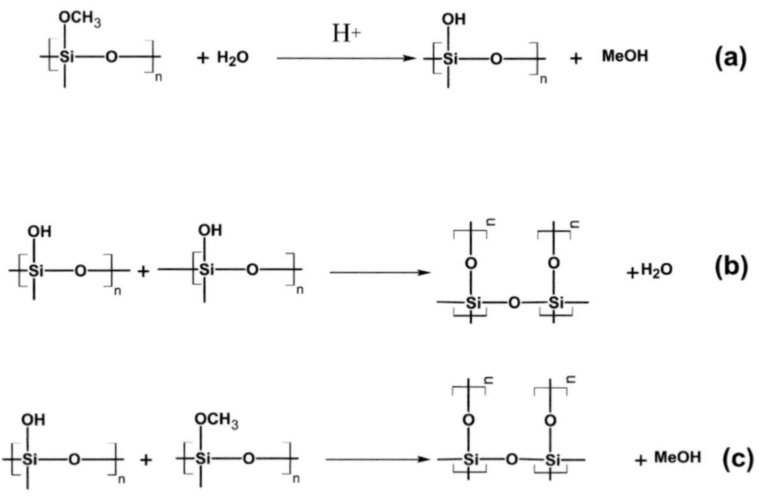

Figure 4.1: Schematics of (a) the Hydrolysis and (b,c) the condensation reactions of alkoxysilane and cross-linking of the polysiloxane.

4.2 Photo-cured polymers

In the coating industry, there are two major mechanisms used to produce photo-cured polymers: cationic and free radical photo-polymerization [79]. These two mechanisms have relatively high speed of curing and are suitable for thin-film applications. Systems based on thiol-enes have also high reactivity and are used in some niche applications. The main advantages of using photo-cured polymers in coatings are [79]:

- Fast curing reaction
- No pot-life (the formulations do not cure in dark before irradiation)
- Low/no volatile organic compounds (VOC)
- Energy and cost saving

In the scientific literature, other mechanisms like cyclo-addition and photo-polymerization based on photo-bases are presented [46, 80]. But most of these mechanisms are very slow and do not reach the performance of the already commercially established systems.

In the following sections, the reaction mechanisms of the reaction and the raw materials of the photo-polymerizations which could be used to produce fast cured dielectrics, are presented. Depending on the application, the presented formulations require some additives like wetting agents, defoamers, colorants and antioxidants. In this work, surface wetting additives are needed since most of the monomers used in photo-curable polymer films have higher surface tension than P3HT. Molecules having one polar group and a non-polar one in their structures, have a preferential orientation at the liquid-air interface: the polar group has an affinity to the polar liquid phase and the non polar group is directed to the air. As a result of this orientation, the surface tension of the mixture is reduced. Surface wetting additives based on silicones and on fluorinated polymers are very efficient in reducing the surface tension.

4.2.1 Cationic systems

Crivello *et al.* and Smith *et al.* showed in parallel in the mid 1970s that diaryliodinium salts and triarylsolphonium salts are efficient in polymerizing olefins, epoxyides, cyclic ethers, acetals and other monomers after exposure to UV-irradiation [81,82]. These salts produce under UV-irradiation cationic active centers which are able to polymerize these monomers. The cationic reaction is insensitive towards air (specially towards oxygen) and produces low shrinkage during the polymerization.

4.2.1.1 Mechanism of the cationic photo-polymerization

The first step of the photo-polymerization is to produce the initiating species by exposing the monomers and photo-initiators mixture to UV-irradiation. A Diaryliodonium salt , for example, is cleaved into two radicals by the UV-irradiation and lead to the production of superacid H^+X^- as shown in following reaction [83]:

$$Ar_2I^+\ X^-\ + RH \quad \xrightarrow{h\nu} \quad ArI\ +\ Ar^\bullet\ +\ R^\bullet\ +\ H^+X^- \tag{4.1}$$

SbF_6^-, PF_6^- and AsF_6^- are widely used as counterions (X^-). Decker *et al.* showed that the cationic reactives have long lifetimes so that their presence could be verified even after stopping the UV-irradiation. That explains the proceeding of the propagation step in dark [84]. In this step, the proton of the superacid is added to the electron rich group (as epoxides). The produced cation can attack further monomers to form a polymer chain as shown in Fig. 4.2. The propagation step is inhibited by moisture (or alcohols) [83], and therefore, the cationic cross-linking of thin films on hydrophilic surfaces is very slow.

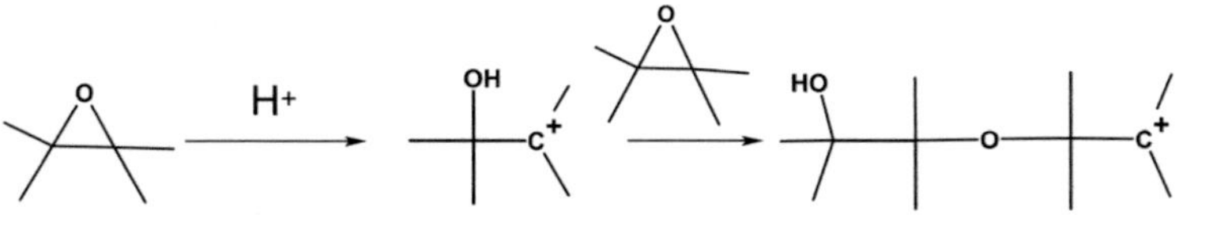

Figure 4.2: Mechanism of the propagation step of the cationic polymerization of epoxides.

4.2.1.2 Raw materials

Many advantages make diaryliodinium salts and triarylsolphonium salts the most commercially used cationic photo-initiators. These salts are stable in room temperature and have low acidity in dark or in daylight, which leads to extended potlife of the formulations. They have also good solubility in epoxies. The counterion of the photo-initiator should have low nucleophilicity to not quench the propagation step of the polymerization [85]. Ordered according to their reactivity, $SbF_6^- > AsF_6^- > PF_6^- > BF_4^-$ could be used for this purpose [85]. Epoxides, especially cycloaliphatic epoxides, are used in the most systems based on cationic photo-curable resins [83]. They are commercially available, produce polymers having high chemical resistance and excellent mechanical properties. In electronics, novolac based resins are used in photo-lithographie, in packaging or as insulator. The phenyl rings in the novolac resin assure high mechanical toughness of the resulting films.

4.2.2 Free radical photo-polymers

Acrylic monomers are the most important monomers used in free radical photopolymerized (FRP) coatings. The preparation and the polymerization of acrylic monomers was first reported in the end of 19th century [86]. In 1940s many patents for the photo-polymerization of unsaturated compounds was granted [87–89]. First industrial products based on photo-curable acrylic systems appeared in the beginning of 1970s [90]. The market of acrylic resins increases since that time because of their advantages compared with physically cured resins. The main applications of acrylic UV-curable systems in conventional electronics are photoresists, printed circuit boards (PCB) and protective coatings. In organic electronics acrylic polymers like poly(methyl methacrylate) (PMMA) are used as insulation layer in OFETs or as passivation layer. Acrylic polymers containing functional groups were also used to get thermally cross-linkable dielectric materials [39]. The use of photocurable acrylic resins is limited to the area of the nanoimprinting techniques were the resins were used as resist [91].

4.2.2.1 Mechanism of free radical photopolymerization

FRP coatings contain unsaturated monomers and photo-initiators that generate free radicals upon UV-irradiation and initiate the polymerization of the monomers. The polymerization of the monomers begins with the addition of one free radical reactive center to a monomer producing a new radical, which can be again added to another monomer. In this way, high molecular weight polymers could be obtained. Fig. 4.3 shows the different steps of the photo-polymerization of acrylic monomers.

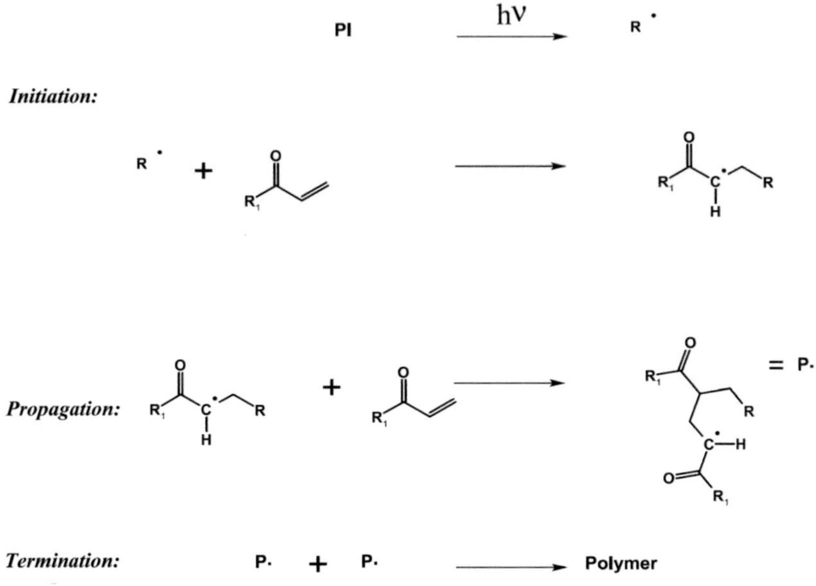

Figure 4.3: Mechanism of the photo-polymerization of acrylates initiated by free radical (PI: Photo-initiator).

4.2.2.2 Raw materials

Monomers

Generally, acrylates (including methacrylates) or unsaturated polyester are used as monomers in FRP coatings. The latter resins have low reactivity and are used only because of their low price. Acrylates have higher reactivity and therefore most of the FRP coatings are based on acrylates. Multifunctional monomers or oligomers should be used when a cross-linked polymer is needed, because monomers with only one functional group lead to linear polymers. Depending on the structure of chemicals used with acrylic acid to produce the acrylic monomers different classes of multifunctional acrylates are defined as shown in Fig. 4.4. The chemical structure of the backbone of the monomers determines the properties of the coatings. So it is important to know the relationship between chemical groups and the properties of the monomers and the resulting films. For example Bisphenol A epoxide based acrylates have generally high reactivity and the resulting films have high hardness and chemical resistance. The good mechanical properties are caused by the phenyl ring in the monomers. To assure film formability on hydrophobic substrates, monomers should exhibit low surface tension, high flexibility and low shrinkage during curing. The surface tension could be reduced by avoiding polar groups and having aliphatic groups which also increase the flexibility of the cured films. The shrinkage is one of the most important issues of the free radical polymerization. Shrinkage could be reduced by using oligomers, monomers with flexible backbone or by reducing the cross-linking density.

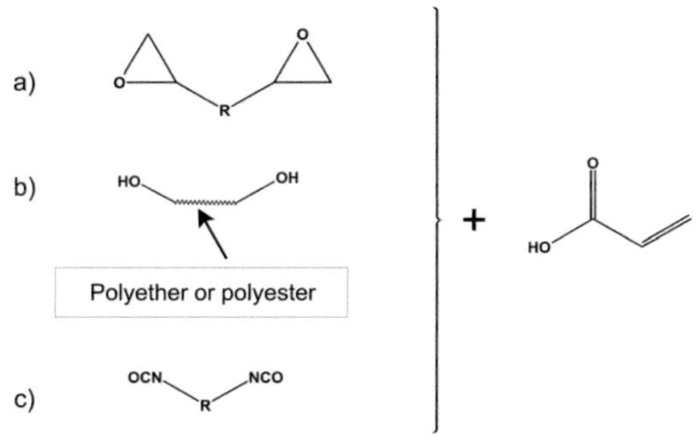

Figure 4.4: General route to produce acrylic multifunctional monomers: Epoxides (a), Polyethers, Polyesters (b) and isocyanates (c) are used to produce acrylic monomers with different properties.

Photo-initiators

A free radical photo-initiating system is a molecule or a mixture of molecules which absorb UV light to generate initiating free radicals. There are two classes of photo-initiators [83]:

- Molecules which undergo Norrish type I cleavage (see Fig. 4.5 a)
- Photoinitiators (type II) generating radicals via hydrogen abstraction (see Fig. 4.5 b).

To obtain thin films ($< 1\mu$m), photoinitiators with good surface curing properties are needed. These are photoinitiators absorbing in UV-C (200 nm - 280 nm), having low oxygen inhibition sensitivity and high reactivity. The oxygen inhibition which is the most important limiting factor of the free radical photopolymerization is more poronounced in thin films [92, 93]. Benzile ketales, α-hydroxyalkyl phenones, α-amino acetophenones as type I photoinitiators and benzophenones as type II photoinitiators are possible candidates to be used to obtain thin films.

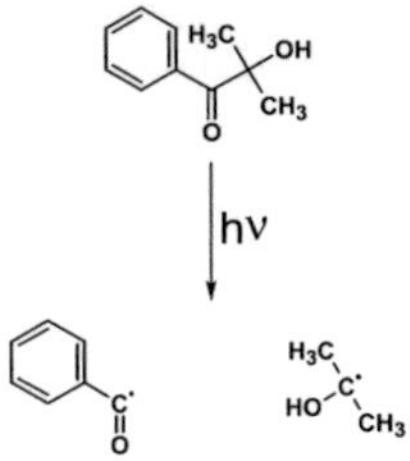

(a)

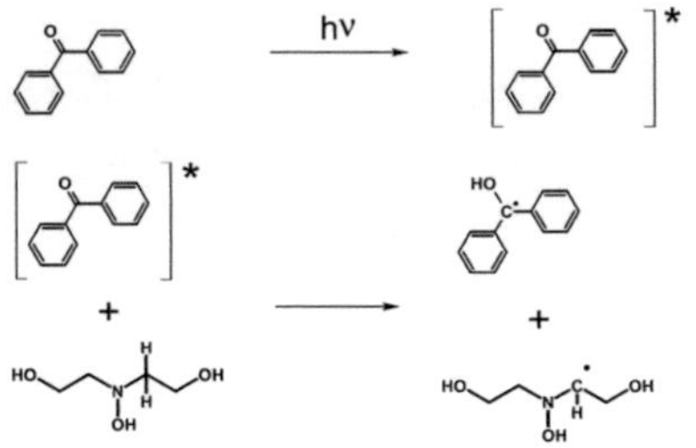

(b)

Figure 4.5: Generation of free radical from (a) 2-hydroxy-2-methyl propiophenone (as a Norrish I photo-initiator) and (b) benzophenone (as a Norrish II photo-initiator) mixed with a tertiary amine under exposure to UV-irradiation.

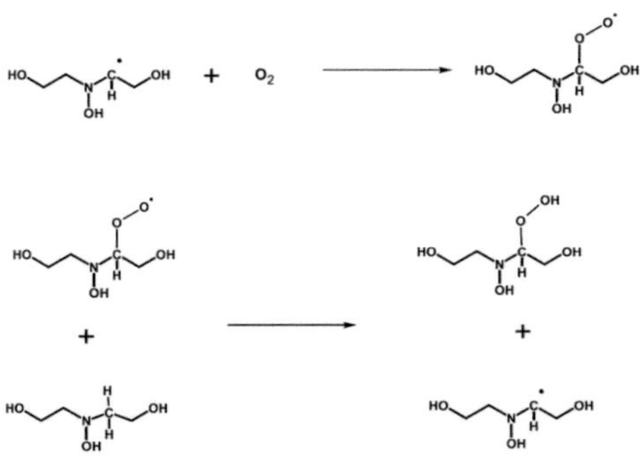

Figure 4.6: Mechanism of the consumption of oxygen by a tertiary amine-radical and the generation of another radical which is able to continue the polymerization reaction.

In the case of photoinitiators type II one hydrogen donor is needed. Generally tertiary amines are used as co-initiators. The mechanism of radical generation using benzophenone-amine system is shown in Fig. 4.5 b. Tertiary amines play another important role in photo-cured polymerized coatings, they consume the dissolved oxygen and thus limit the oxygen inhibition. As shown in Fig. 4.6 the amine scavenges the oxygen and produces a peroxy radical which abstracts a hydrogen from another amine forming one reactive amine radical which can consume another oxygen or initiate the polymerization [94].

4.2.3 Thiol-ene polymers

Thiol-ene reaction was investigated in the 1930s [95]. It was first studied as a monoaddition reaction. In 1948, Marvel *et al.* showed that diolefins and dimercaptans can undergo an addition reaction to produce a polyalkylene sulfide [96]. The thiol-ene reaction can be initiated by free radicals, which could be generated by using free-radical photo-initiators or by using peroxides. Thiol-ene photo-polymerization is efficient in producing highly cross-linked polymer networks in air since it exhibits no or low sensitivity to oxygen [97]. Other advantages for these mixtures include fast reaction, low shrinkage and flexibility of the produced films [97]. The main applications of thiol-ene polymers in electronics are in the coating of printed circuit boards, electrical wire and optical fibers [94].

4.2.3.1 Mechanism of the thiol-ene reaction

The thiol-ene reaction is initiated by the generation of thiyl radicals. The latter radicals could be produced by exposure of the thiols to UV-irradiation or by chain-transfer process with other radicals, e.g. cleaved photo-initiators. The produced thiyl radicals undergo a step-growth addition reaction with the olefins. Depending on the chemical nature of the olefin, homopolymerization of the olefin can occur and competes with the thiol-ene reaction. This was observed when acrylic monomers were used [97]. In the other cases, where homopolymerization is negligible, it is important to preserve a stoichiometric ratio of 1:1 of the thiols and the enes. The propagation of the reaction is maintained by a chain transfer reaction. That means the radical produced by the addition of the thiyl to the olefin can abstract a hydrogen from a further thiol to produce another thiyl radical. The termination of the reaction occurs by radical-radical coupling. The mechanisms of the initiation, the step-growth addition and the chain transfer reaction are shown in Fig. 4.7.

The thiol-ene reaction is not inhibited by oxygen because the peroxy radical can abstract a hydrogen from a thiol and produce a thiyl radical which preserve the propagation of the reaction (see Fig. 4.8) [98].

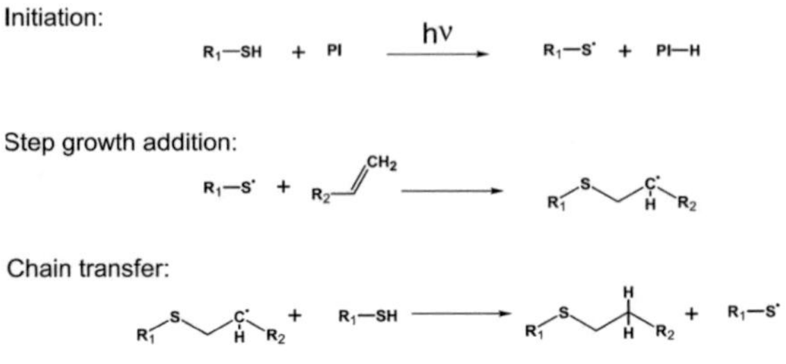

Figure 4.7: The mechanism of thiol-ene photo-polymerization (PI: Photo-initiator).

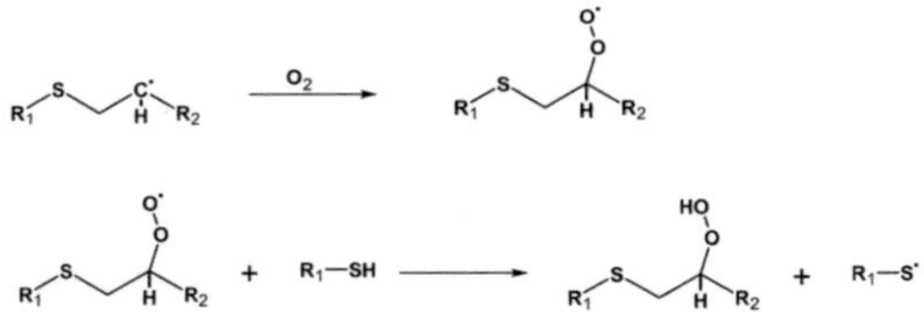

Figure 4.8: Reaction of the thiyl radical with oxygen and the chain transfer reaction with a further thiol.

4.2.3.2 Raw materials

A thiol-ene is a mixture of multifunctional thiols (mercaptan) and multifunctional enes (e.g. acrylate, vinyl-ether, allyl) [94]. There are two types of multifunctional thiols which are commercially available: thiol glycolate esters and thiol propionate esters (see Fig. 4.9). These monomers are produced by esterification of thioglycolic acid or 3-mercaptopropionic acid with a polyol. The two classes of thiols have enhanced reactivity with ene groups because of the weakening of the sulfur-hydrogen bond due to the interaction with the carbonyl group [94, 99]. Some of the commercially available thiols are shown in Fig. 4.9.

Any class of enes (e.g. norbornene, vinyl ether, (meth)-acrylates, allyl ether and many others) can be involved in a reaction with thiols [97]. The reactivities of norbornenes and vinyl ethers with thiols is higher than the other monomers [97]. The same criteria for choosing the right monomers for the acrylates are also valid for the selection of the ene for the thiol-ene polymers. Furthermore, the photo-initiators used in acrylic polymers could be also used in thiol-ene formulations.

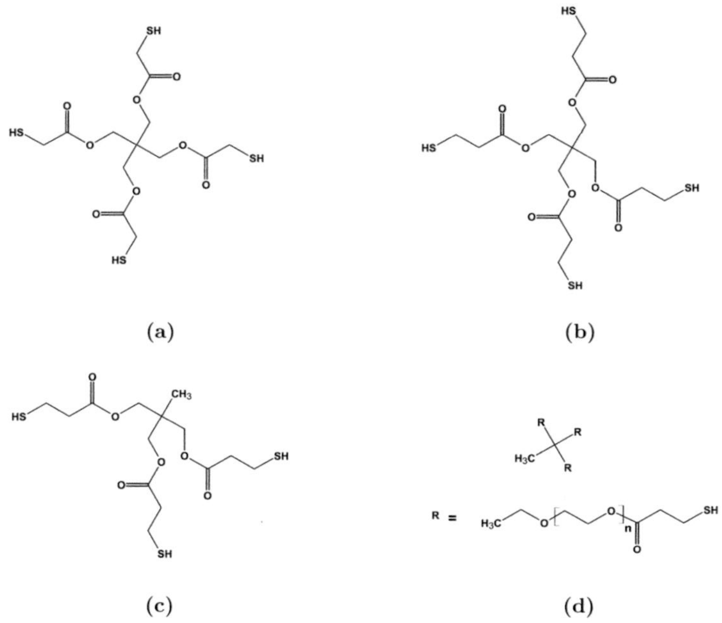

Figure 4.9: Structure of some commercially available thiols: (a) Pentaerythritol Tetramercaptoacetate (PETMA) (b) Pentaerythritol Tetra-3-mercaptopropionate (PETMP) (c) Trimethylolpropane Tri(3-mercaptopropionate) (TMPMP) (d) Ethoxilated-Trimethylolpropan Tri-3-Mercaptopropionate (ETTMP)

5. Dielectric materials based on linear polymers

In this chapter, first experiments with dielectric materials based on polymers are shown and interpreted. The aim of these experiments is not to cover all the polymer classes which are suitable for OFETs. Since, as shown in chapter 1, many linear polymer classes are already reported to be suitable as insulators in OFETs. The objective of the investigation of linear polymers is to have one "standard"-behavior of OFETs to be used as reference in the further work. Styrenic polymers with different side groups are investigated: polystyrene (PS), poly(α-methylstyrene) (PαMS), poly(4-methylstyrene) (P4MS),poly(4-tertbutyl-styrene) (P4tbS) were used as "neutral" polymers. Poly(4-vinylpyridine) (P4VPy) and poly(2-vinylpyridine) (P2VPy) are lewis bases and poly(4-vinylphenol) (P4VP) has an acidic character. Suitable solvents for the polymers were selected. Then, the dielectric properties of films and the performance of OFETs based on these polymers were measured.

5.1 Materials and their properties

This section presents the investigated styrenic polymers and their properties. Polystyrene (PS), poly(α-methylstyrene) (PαMS), poly(4-methylstyrene) (P4MS), poly(4-tert-butylstyrene) (P4tbS), poly(4-vinylpyridine) (P4VPy) and poly(2-vinylpyridine) (P2VPy) were purchased from Scientific Polymer Products. Poly(4-vinylphenol) (P4VP) was purchased from Sigma Aldrich. PS, P4MS, P4tbS and P4VP are soluble in ketones and esters and therefore suitable for top-gate P3HT based OFETs. P4VPy and P2VPy could be also used in top-gate P3HT-based OFETs since P4VPy is soluble in lower alcohols (like ethanol, n-propanol) and P2VPy is soluble in ketones and in lower alcohols. However, PαMS is only soluble in solvents which dissolve

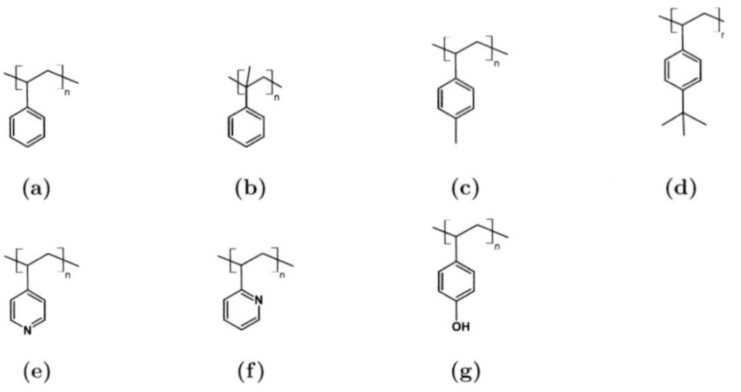

Figure 5.1: Chemical structure of (a) polystyrene (PS) (b) poly(α-methylstyrene) (PαMS) (c) poly(4-methylstyrene) (P4MS) (d) poly(4-tert-butylstyrene) (P4tbS) (e) poly(4-vinylpyridine) (P4VPy) (f) poly(2-vinylpyridine) (P2VPy) (g) poly(4-vinylphenol) (P4VP)

P3HT (Toluene, xylene, chloroform). For this reason, PαMS was excluded from further investigations.
PS, P4MS, P4tbS and P4VP were dissolved in methyl-ethylketone (MEK, B.p. 80 °C), isopropylacetate (B.p. 89 °C) and n-butylacetate (B.p. 127 °C). Two solutions of P2VPy in n-propanol and in a mixture of MEK and n-propanol were prepared. P4VPy was dissolved in n-propanol. After spin-coating the solutions on glass substrates, the surfaces of the films were scanned with a profilometer (DEKTAK 3030, Veeco). Fig. 5.2 shows that the roughness of films made with PS increases when MEK or iso-propylacetate were used as solvent. The use of the solvent with higher boiling point reduces the roughness of the films. The same effect was seen with P4MS, P4tbS and P4VP. P2VPy and P4VPy solutions in n-propanol produced smooth films.

The polymers were dissolved in the selected solvents and then used as insulators in MIM-structures. Fig. 5.3 shows the current density flowing through the MIM-structures in dependence of the electrical field. These measurements showed that PS, P4MS and P4tbS are better insulating than the more polar polymers (P4VP, P4VPy and P2VPy). The current density was lower than 10^{-9} A/cm^2 at an electrical field of 0.7 MV/cm. The leakage current was approximately 6 and 30 times higher when P2VPy and P4VPy were used, respectively. PS, P4MS and P4tbS have a low dielectric constant which is equal to approximately 2.3. On the other hand,

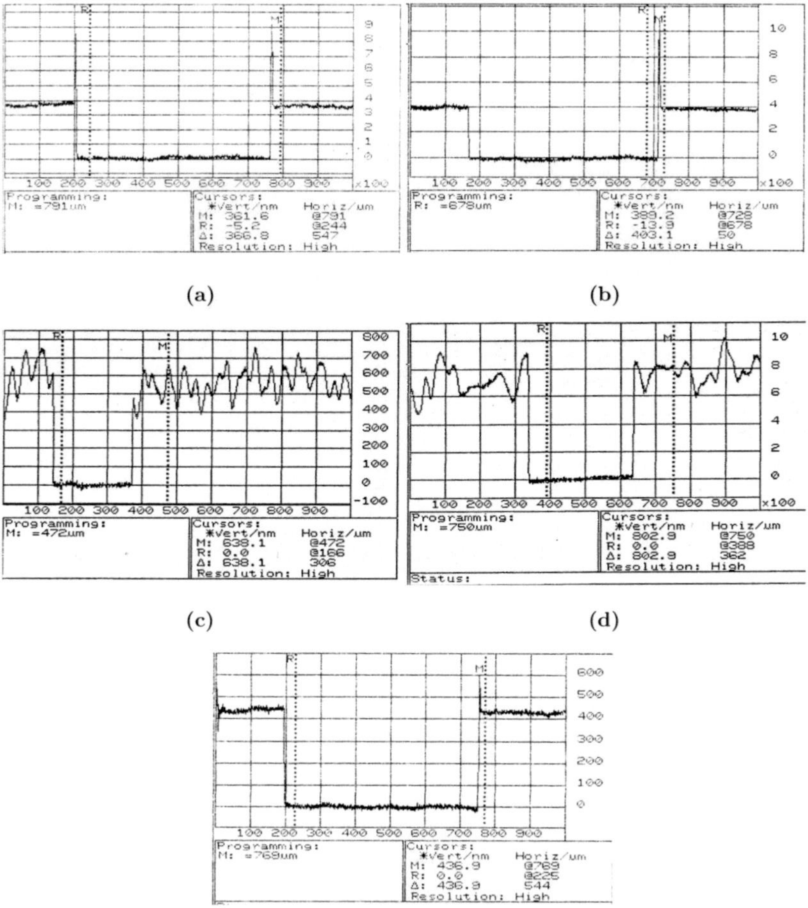

(e)

Figure 5.2: Surface topography of films made with (a) P2VPy in n-propanol (b) P4VPy in n-propanol (c) PS in MEK (d) PS in iso-propyl acetate (e) PS in n-butyl acetate solutions [14].

P4VP, P4VPy and P2VPy have higher dielectric constant because of their polar groups. Volume resistivity of the films were calculated and summarized with the measured dielectric constants in table 5.1.

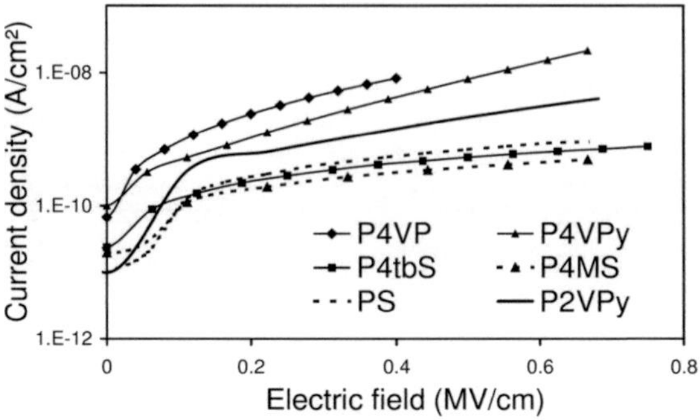

Figure 5.3: Current density vs. electrical field in MIM-structures made with styrenic polymers [14].

Polymer	ρ [Ω cm]	ϵ
PS	7.3×10^{14}	2.4
P4MS	1.4×10^{15}	2.3
P4tbS	9.7×10^{14}	2.3
P4VPy	3.1×10^{13}	3.7
P2VPy	1.7×10^{14}	4.5
P4VP	4.8×10^{13}	4.8

Table 5.1: Volume resistivity and dielectric constant of styrenic polymers [14].

5.2 Transistors

Top-gate transistors based on styrenic polymers were fabricated. The polymers were dissolved in the selected solvents and then spin-coated on P3HT, to get films with thicknesses between 300 and 400 nm. Transfer characteristics of these transistors are shown in Fig. 5.4. The logarithmic plot of the transfer characteristics shows that the on-currents of the transistors made with PS, P4MS, P4tbS, P2VPy and P4VPy are almost equal. However, the off-currents and the threshold-voltages of these transistors were different. The on and the off currents of the transistor made with P4VP was many order of magnitude higher than the other transistors. Because its threshold-voltage is around +40 V, the on current of the transistor made with P4VP at -40 V is higher than transistors having their threshold-voltage around 0 V. The off current was high because of the high leakage current, which was between 0.3 μA and 2 μA. The root squared plots of the transfer characteristics were not linear and therefore, the use of the equation 2.6 was necessary to extract the values of the threshold-voltages and the field-effect mobilities. Table 5.2 contains the most important parameters of the transistors made with the styrenic polymers. The threshold voltages of all the transistors were positive, which means that at $V_{gs} = 0\ V$, the current flowing between the source and the drain of a transistor is higher than its off-current. The calculated field-effect mobilities of the transistors made with the investigated polymers except P4VPy were in the range between 1 and 2 $cm^2\ V^{-1}\ s^{-1}$. P4btS with its on/off ratio of 10^4 and its field-effect mobility of $2.1 \times 10^{-2}\ cm^2\ V^{-1}\ s^{-1}$ could be used as reference for the investigation of the cross-linked polymers.

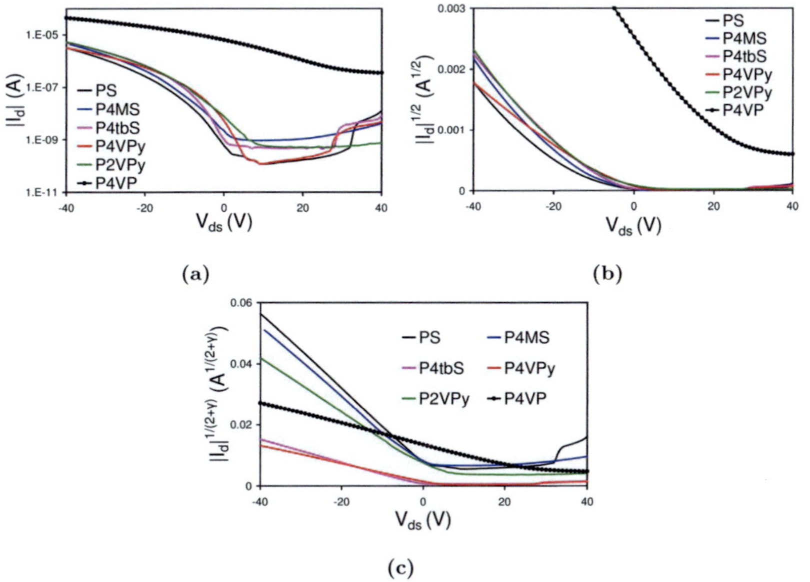

Figure 5.4: (a) Logarithmic and (b) root squared transfer characteristics of transistors made with styrenic polymers. (c) Plot of measured current raised to the power $1/(2+\gamma)$ [14].

Polymer	ϵ_r	γ	μ @ -40 V [$cm^2\ V^{-1}\ s^{-1}$]	On/off	SS [V/dec]	N_{SS} [cm^{-2}]	V_{th} [V]
PS	2.4	2.4	1.7×10^{-2}	2.7×10^4	5.25	2.6×10^{12}	5.9
P4MS	2.3	2.1	2×10^{-2}	5.1×10^3	5.9	3.3×10^{12}	5.4
P4tbS	2.3	0.91	2.1×10^{-2}	1×10^4	3.8	2.1×10^{12}	0.1
P4VPy	3.7	0.92	4.2×10^{-3}	2.6×10^4	3.3	2.7×10^{12}	4.5
P2VPy	4.5	1.8	1.1×10^{-2}	1×10^4	11.4	1.2×10^{13}	7.2
P4VP	4.8	0.78	2.8×10^{-2}	1.2×10^2	25	2.3×10^{13}	39.5

Table 5.2: Electrical properties of top-gate transistors fabricated with styrenic polymers as dielectric materials.

5.3 Conclusion

The "neutral" polystyrenic polymers, which have low dielectric constants, have high volume resistivity. The transistors made with these polymers have on/off-ratios of about 10^4 and free charge carrier mobilities of about $10^{-2}\ cm^2\ V^{-1}\ s^{-1}$. The transistors based on polyvinylpyridines (P2VPy and P4VPy) show similar characteristics. Only the transistors based on P4VP are suffering from high leakage currents and high positive V_{th}.

6. Dielectric materials based on thermally curable polymers

In this chapter two approaches to reduce the time of curing of cross-linkable dielectrics are presented. The first one is to use polymers with reactive groups assuring the cross-linking of their mixtures. P2VPy, a polymer having good properties as dielectric layer in OFETs (see chapter 5), could be cross-linked using a novolac resin, which contains epoxide groups. The second material system, is based on a polysiloxane with methoxy side-groups. A high reactive catalyst based on an organo-titanium compound is used to accelerate the curing process of the dielectric.

6.1 Cross-linked Poly(2-vinylpyridine)

6.1.1 Curing process

A mixture of MEK and 1-propanol was used as solvent for the blend of P2VPy and DEN.425, because the novolac resin is not sufficiently soluble in 1-propanol. Thin films of the mixture were produced by spin-coating and curing at 140 °C. The curing process was accompanied with the apparition of an orange color. Therefore films with different thicknesses were characterized with UV-Vis spectroscopy before and after the cross-linking process. Fig. 6.1 shows the UV-Vis absorption spectra of P2VPy-DEN.425 films before and after curing at 140 °C. The absorption between 300-500 nm increased (better visible in the case of the thick film) and thus the films became orange. The band at 280 nm decreased during the curing (better visible in the case of the thin film). The changes of the absorption spectra are negligible after 30 min of curing.
The FTIR spectrum of the mixture before and after the cross-linking reaction gives more details about the reaction mechanism. The bands of the monosubstituted epoxy and of the CH_2 of the epoxy group at 840 cm^{-1} and 916 cm^{-1} [100] respec-

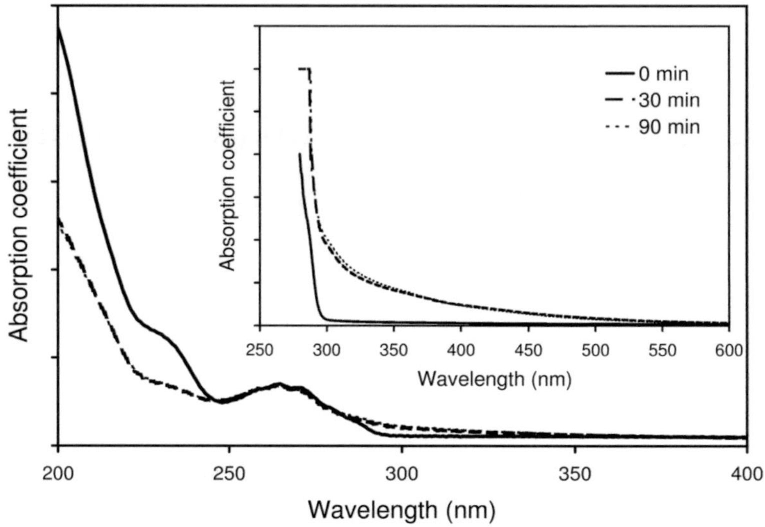

Figure 6.1: UV-VIS absorption spectrum of a thin layer and a thick layer (inset) of a mixture of P2VPy and DEN.425 during the curing process. The thickness of the layers was 20 nm and 4 μm respectively.

tively disappeared during the reaction. A new band around 3300 cm^{-1} [100], which is assigned to the OH group, is appeared. The FTIR confirms also that P2VP is involved in the reaction, since it shows an increase of the intensity of the bands at 1645 cm^{-1}, 1650 cm^{-1} and 1656 cm^{-1} [100], which could be ascribed to the carbonyls of pyridone and the unconjucated carbon-carbon double bonds. A possible mechanism of the reaction is shown in Fig. 6.3. Both measurements (FTIR and UV-Vis) showed that curing the films for more than 30 minutes did not increase the grade of the cross-linking of the polymer.

The produced cross-linked polymer is insoluble in common solvents like acetone, toluene and n-Butylacetate. The volume resistivities of films made with 10:1 and 2:1 mixtures of P2VPy and DEN.425 were $0.55 \times 10^{15}\Omega\ cm$ and $0.37 \times 10^{15}\Omega\ cm$ at 1 MV/cm respectively. These values were slightly higher than those of pure P2VP ($0.14 \times 10^{15}\Omega\ cm$).

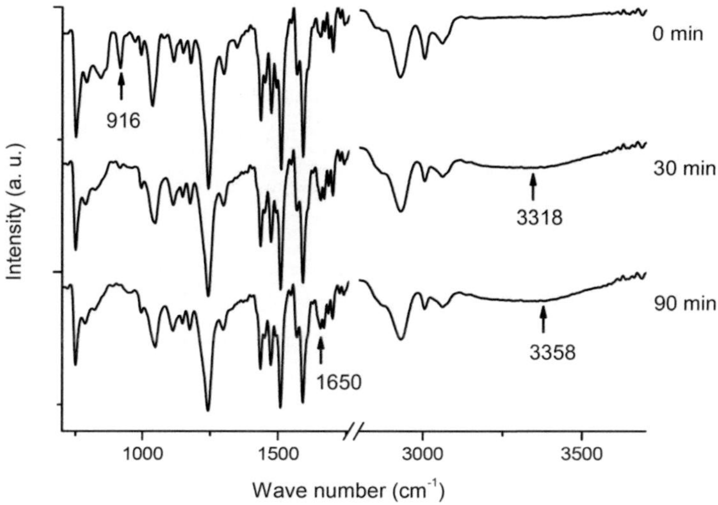

Figure 6.2: FTIR spectrum of a film made with a mixture of P2VPy and DEN.425 during the curing process.

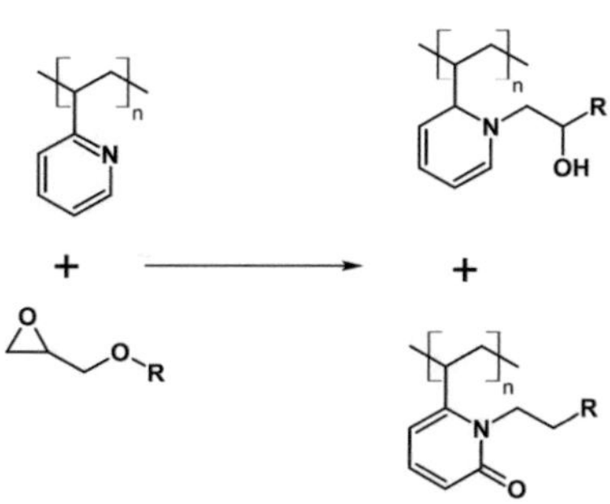

Figure 6.3: Cross-linking reaction between P2VPy and DEN.425.

6.1.2 CV-measurement

The addition of the novolac resin and the curing process have only a negligible effect on the semiconductor surface as shown in Fig. 6.4. The doping concentration calculated from the $1/C^2$-V slope is almost identical in all the measured MIS-structures and was equal to around 2 $\times 10^{16}$ cm^{-3}. All the devices made with cross-linked P2VPy exhibit also negative flat-band voltages between -6 V and -7.3 V similar to devices made with pure P2VPy.

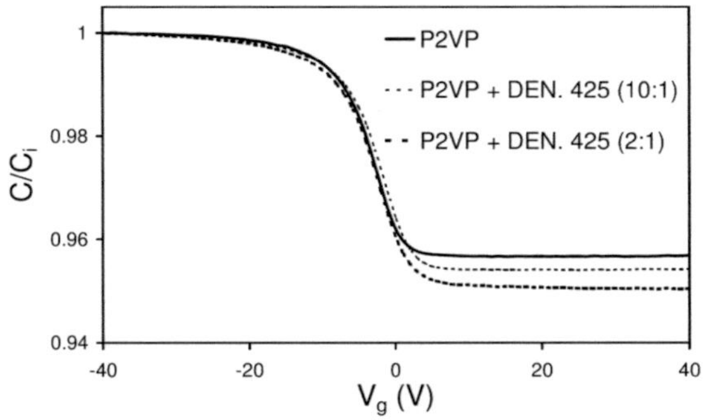

Figure 6.4: CV-characteristics of MIS-structures made with pure and cross-linked P2VPy.

The V_{fb} shift is caused by fixed charge N_f(or traps with life time longer then the measurement time) in the dielectric layer with a concentration between 1.3 $\times 10^{11}$ cm^{-2} and 1.47 $\times 10^{11}$ cm^{-2}. The pyridine group in the P2VPy is a lewis base, which means it has the readiness to give one electron of its lone pairs to its environment. Therefore, P2VPy can produce donor-like traps and becomes positively charged at the interface with P3HT. These results show that the addition of DEN.425, the curing process and its products do not have any significant effect on the interface between the semiconductor and the dielectric material. The CV-characteristics of all the structures showed also no hysteresis in nitrogen and in air. Table 7.2 summarizes the parameters extracted from the CV-measurements.

	P2VP	P2VP:DEN.425 10:1	P2VP:DEN.425 2:1
N_A (10^{16} cm^{-3})	1.8	2	2
V_{fb} (V)	-6.5	-7.3	-6
N_f (10^{11} cm^{-2})	1.36	1.47	1.3

Table 6.1: Doping concentration, flat-band voltage and concentration of fixed charges extracted from the CV-characteristics of MIS-structures made with pure and cross-linked P2VPy.

6.1.3 Transistor

Transistors made with P2VPy and DEN.425 mixtures show on/off ratios approximately equal to 10^5. The thickness of the dielectric films used to fabricate the transistors were 350 nm and 700 nm for 2:1 and 10:1 mixtures of P2VPy and DEN.425 respectively. The difference of the thickness of the films could explain why the transistors made with the higher concentration of DEN.425 have higher on and off-currents. As shown in Fig. 6.5 (b), the transfer characteristics could be well fit with equation 2.6. The V_{fb} shift was negative and low in both transistors (-2.1 and -4.5 V), which means that at $V_{gs} = 0\ V$ the transistors were in the off-state. Transistors made with 2:1 mixture have higher free charge carriers mobilities and lower on/off ratios than their counterparts made with 10:1 mixture. It has also a low sub-threshold swing which was equal to 0.7 V/dec. The properties of the transistors are summarized in Table 6.2. Compared with the styrenic polymers, the transistors made with cross-linked P2VPy have higher on/off-ratios but lower field-effect mobilities.

Polymer	ϵ_r	γ	μ @ -40 V $[cm^2\ V^{-1}\ s^{-1}]$	On/off	SS $[V/dec]$	N_{SS} $[cm^{-2}]$	V_{th} $[V]$
P2VPy + DEN.425 (2:1)	5.9	1.3	1.3×10^{-3}	8.9×10^4	0.7	1×10^{12}	-4.5
P2VPy + DEN.425 (10:1)	5.9	1.7	5×10^{-4}	1.8×10^5	1.5	2.2×10^{12}	-2.1

Table 6.2: Electrical properties of top-gate transistors fabricated with P2VPy and DEN.425 mixtures as dielectric material.

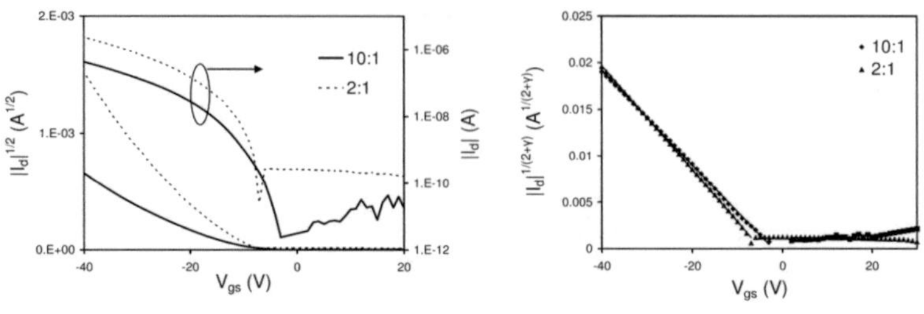

(a) (b)

Figure 6.5: (a) Root squared and logarithmic transfer characteristics of transistors made with P2VPy and DEN.425, (b)plot of measured current raised to the power $1/(2+\gamma)$.

6.2 Dielectric based on Polysiloxane

6.2.1 Raw Materials

Silres MSE 100, which is a methoxy functionalized polymethoxysiloxane, was supplied by Wacker AG. Its curing reaction occurs in the presence of humidity and is catalyzed by acids or bases. The producer of Silres MSE 100 recommend the use of titanium compounds as catalyst. Titanium diisopropoxide bis (acetylacetonate) (TAA) (Dorf Ketal) is a lewis acid which initiates, in the presence of humidity, the cross-linking process of methoxy functionlaized polysiloxanes. Both materials are in liquid form and they are mixable with alcohols and esters (e.g. n-butylacetate). The amount of used solvent could be reduced in this systems since the two materials are liquid and because of the possibility of using other polysiloxanes with lower viscosity.

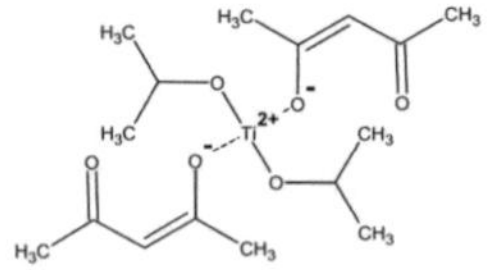

Figure 6.6: Structure of titanium diisopropoxide bis (acetylacetonate) (TAA).

6.2.2 Curing process

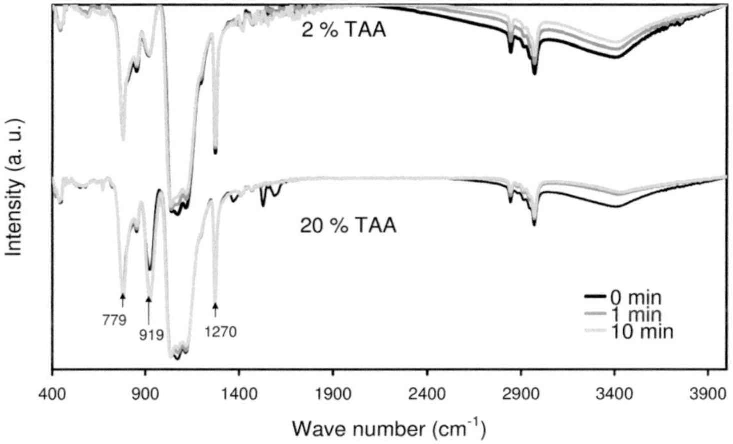

Figure 6.7: FTIR spectrum of films made with a methoxy-functionalized polysiloxane during the curing process with two different concentrations of TAA.

The polysiloxane is hydrolyzed in presence of TAA and humidity, and then condensed by producing Si-O-Si and Si-O-Ti bonds. TAA reacts also with humidity and produces titanium dioxide. The FTIR study of the spin-coated films shows that Ti-O-Si (919 cm^{-1}) [100] bonds are already established after spin-coating even before the treatment with high temperature. After one minute of curing at 120 °C no Ti-O-Si are further produced. The monitoring of the Si-O-Si (980 cm^{-1} - 1080 cm^{-1}) [100] bonds is more difficult since the backbone of the polysiloxane consists of Si-O-Si bonds.

The curing process of the polysiloxane could be characterized by measuring the dielectric properties after different curing times. For this purpose, films of polysiloxane were prepared with different concentrations of TAA and cured at 120 °C. After 30 min of curing, the temperature was increased to 150 °C. Fig. 6.8 (a) shows that during the cross-linking reaction, the dielectric constant and loss are decreasing. This tendency is more significant for films with low concentration of TAA. This could be correlated with the fact that these films take also more time to become tack-free. Fig. 6.8 (b) shows that the dielectric properties of films cured for 5 min are strongly dependent on the concentration of TAA. It shows also that low dielectric loss is only obtainable after 60 min of curing if low concentration of TAA is needed. These results could be confirmed if the volume resistivity of the films is characterized (see Fig. 6.9). A volume resistivity of $2 \times 10^{14}\ \Omega\ cm$ is only reachable after 60 min of curing. When the films are cured only for 5 min the volume resistivity is highly dependent on the TAA concentration, and only at high concentration of TAA the films were fully cured. At low concentration of TAA longer time of curing was necessary to get cure the films and to get high volume resistivities.

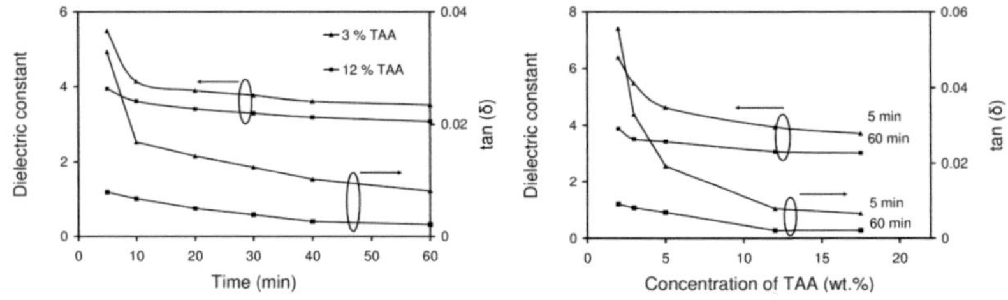

(a) (b)

Figure 6.8: (a) Dielectric constant and loss of films made with the polysiloxane cured with two different concentrations of TAA vs. curing time.(b) Dielectric constant and loss of films made with the polysiloxane cured for 5 and 60 minutes vs. TAA concentration [15].

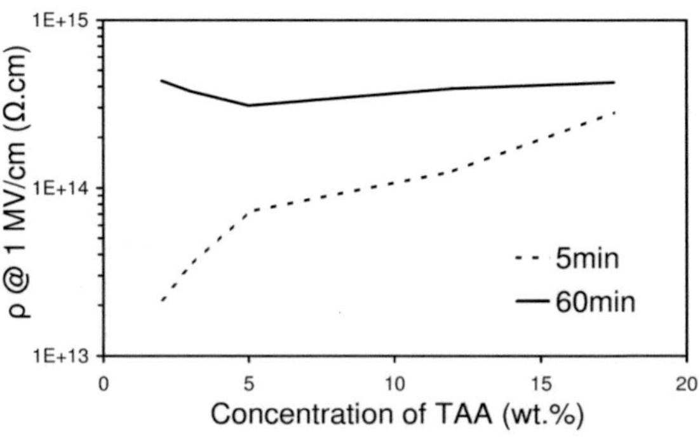

Figure 6.9: Volume resistivity of films made with the polysiloxane cured for 5 and 60 minutes vs. TAA concentration [15].

6.2.3 CV-measurement

During the curing process, P3HT comes in contact with TAA, non reacted methoxy groups and with other reactive groups e.g. silanol. To determine the effect of this curing mechanism on the semiconductor, MIS-structures containing polysiloxane with different concentrations of TAA were fabricated. As shown in Fig. 6.10, V_{fb} becomes more positive with increasing concentration of TAA. Even when only 1 wt.% of TAA is used, V_{fb} has value of about 10 V. This V_{fb}-shift is caused by the acidicity of TAA. TAA is a lewis acid, that means it can accept electrons from its environment, therefore lewis acids are used as dopant for organic semiconductor [101]. If these entities comes in contact with the surface of P3HT they could generate acceptor-like traps, so they cause a positive V_{fb} shift in CV-characteristics [102].

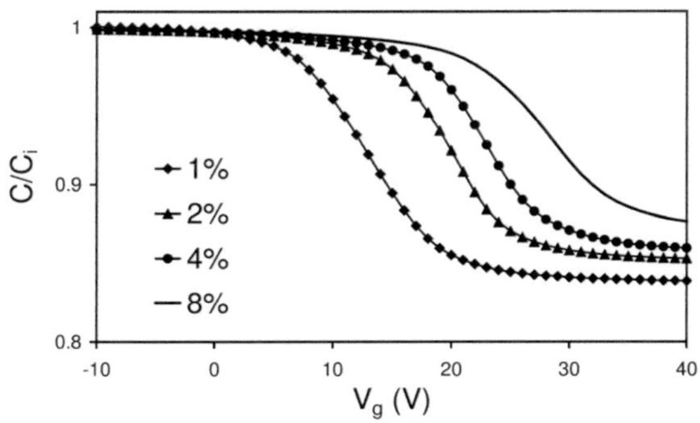

Figure 6.10: CV-characteristics of MIS-structures made with polysiloxane cured with different TAA concentrations [15].

The curing of Silres using bases was investigated to avoid the V_{fb}-shift caused by acidic catalysts. The use of alkali bases (KOH, NaOH ...) is excluded because they generate ions in the dielectric. First experiments with amines showed that the curing process of Silres using bases is much slower than if acids were used. Silres needs at least 12 h of curing at RT when polyether diamine (secondary amine) is used as catalyst. The films could not be cured if they were directly treated with heat (120 °C). To speed up the curing of the films, bases with higher pK_a are needed. 1,8-Diazabicycloundec-7-ene (DBU) ($pK_a = 12$) belongs to the class of amidine and is a strong organic base (see Fig. 6.11). Its effect on the V_{fb} is shown in Fig. 6.12. MIS-structures were made with silres blended with different concentrations of DBU and cured for 15 min in RT and then for 45 min at 150 °C. Films containing 8 wt.% of DBU produced a V_{fb}-shift of -5.9 V, which is similar to the shift caused by P2VPy and its mixtures.

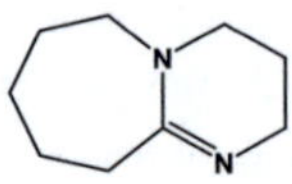

Figure 6.11: Structure of 1,8-Diazabicycloundec-7-ene (DBU).

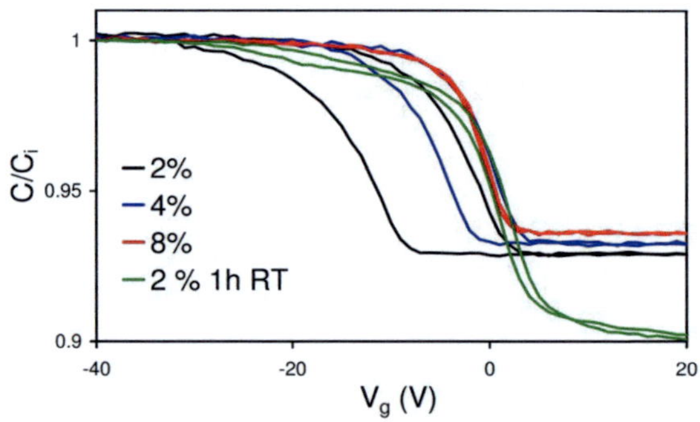

Figure 6.12: CV-characteristics of MIS-structures made with polysiloxane cured with 2, 4 and 8 wt.% of DBU cured for 15 min at room temperature and than for 45 min at 150 °C. The green curve correspond to a device made with polysiloxane cured with 2 wt.% of DBU cured for 60 min at room temperature and than for 45 min at 150 °C.

DBU is also one lewis base and can generate also donor-like trap. Since the V_{fb}-shift is positive (unlike the MIS-structure produced using TAA as catalyst), it is expected that the transistors fabricated by using DBU as catalyst would have low drain-source current at $V_{gs} = 0\ V$. The films with lower concentrations of DBU (2 and 4 wt.%) produced hysteresis in the CV-characteristics. When the films were cured longer (1 h in RT and then 45 min at 150 °C), the hysteresis disappeared and the V_{fb}-shift is similar to case of films with 8 wt.% of DBU. As a result of the dependance of the V_{fb} on the catalyst used to cure the films, the system based on polysiloxanes with TAA and DBU could be used on applications where V_{th} engineering is required, as in the case of the depletion-load inverters where a depletion- and an enhancement-mode transistors are required [18].

6.2.4 Transistors

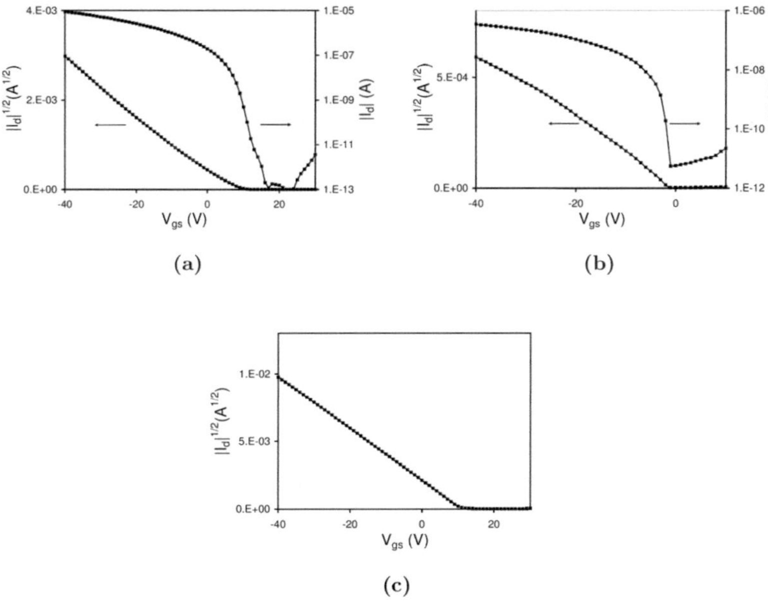

Figure 6.13: Root squared and logarithmic transfer characteristics of transistors made with Silres cured with TAA [15] (a) and with DBU(b). Plot of the drain-current raised to the power $1/(2+\gamma)$ of a transistor made with Silres and TAA [15].

The CV-characterization of the interface between P3HT and polysiloxane cured with different catalysts showed that BDU is more suitable for organic transistors. The curing time of polysiloxane could not be reduced. At least one hour was necessary to obtain films having high volume resistivity and producing low V_{fb} shift. Transistors made with polysiloxane cured with 2 wt.% of TAA, have high on/off ratios (approximately 10^7) and a high positive $V_{th} = +10$ V (see Fig. 6.13 (a)). The transfer characteristic could be fit very well with the equation 2.6.(see Fig. 6.13 (c)). The fitting parameter γ and the mobility at $V_{gs} = -40\ V$ were equal to 0.51 and $4.7 \times 10^{-3}\ cm^2V^{-1}s^{-1}$ respectively. The sub-threshold swing was 1.57 V/dec corresponding to a trap density of about $9.6 \times 10^{11}\ cm^{-2}$. In the other hand, transistors cured with 8 wt.% of DBU, have lower sub-threshold swing (0.65 V/dec), which corresponds to a trap density of about $2 \times 10^{11}\ cm^{-2}$. The transistors have low V_{fb} shift (-1 V), which means a very low current is flowing from drain to source at $V_{gs} = 0$ V. However, the on/off ratios were approximately two order of magnitude lower than transistors cured with TAA. The transfer characteristics of transistors made with polysiloxane cured with DBU could not be fit with equation 2.6. and therefore the mobility was calculated from the root squared transfer characteristic. The mobility was approximately one order of magnitude lower than their counterparts cured with TAA. Both transistors have lower field-effect mobilities and higher on/off ratios than the transistors made with P4tbS. Table 6.3 summarizes the electrical properties of the transistors made with poylsiloxane cured with 2 wt.% of TAA and 8 wt.% of DBU.

Polymer	ϵ_r	γ	μ @ -40 V $[cm^2\ V^{-1}\ s^{-1}]$	On/off	SS $[V/dec]$	N_{SS} $[cm^{-2}]$	V_{th} $[V]$
Silres + TAA	3	0.51	4.7×10^{-3}	8.6×10^6	1.57	9.6×10^{11}	10.1
Silres + DBU	3	-	4.4×10^{-4}	6.4×10^4	0.65	2×10^{11}	-1

Table 6.3: Electrical properties of top-gate transistors fabricated with polysiloxane cured with TAA and DBU as dielectric material.

6.3 Conclusion

In this chapter, cross-linkable dielectric materials based on two different polymers were presented. The first system was based on P2VPy which was cross-linked with a novolac epoxy resin. The second one was based on methoxy-functionalized polysiloxane which was cured with a lewis acid or with a lewis base. Both systems need a curing time higher than 30 min to get films having high volume resistivity and good chemical resistance to common solvents. CV-measurements showed that devices made with a dielectric material containing a lewis base (P2VPy or polysiloxane cured with DBU) have a negative V_{fb} shift. Devices containing a lewis acid in the dielectric layer have a much higher V_{fb} shift in the positive direction. In the case of polysiloxane, it was only possible to reduce the time of curing by increasing the TAA concentration, which produced high positive V_{th}. Although the transistors made with these materials show very encouraging properties, the long curing time is a drawback for these systems to be used in high capacity processes for the production of organic transistors.

7. Dielectric materials based on photocurable polymers

Photo-curable polymers are advantageous for producing printable organic transistors, because of their fast cure and low/no volatile organic content (VOC). In this chapter, three different mechanisms to produce photocurable dielectrics for top-gate transistors are presented. CV-measurements give more information about the effect of the UV-irradiation and the reactive substances on the semiconductor-insulator interface.

7.1 Cationically polymerized dielectrics

7.1.1 Raw materials

The first photo-curable polymer investigated in this work is based on a cationic system composed by SilForce UV9390C as photo-acid-generator (PAG), DEN.425 and P4VP as cross-linkable compounds. SilForce UV9390C is a mixture of bis(4-dodecylphenyl)-iodonium hexafluoroantimonate, a photosensitizer and a reactive diluent. It was supplied by Momentive Performance Materials Inc. and was used as PAG. It was chosen because of its high absorption at 255 nm. It was blended with DEN.425 (novolac epoxy resin) and poly(4-vinylphenol) (PVP). DEN.425 was supplied by Dow Chemical Company and PVP was supplied by DuPont. N-Butylacetate was purchased from Carl Roth and used as solvent. Efka 3600, which is a polymeric fluorocarbon compound, was supplied by BASF and was used as a wetting agent. All the chemicals were used without purification.

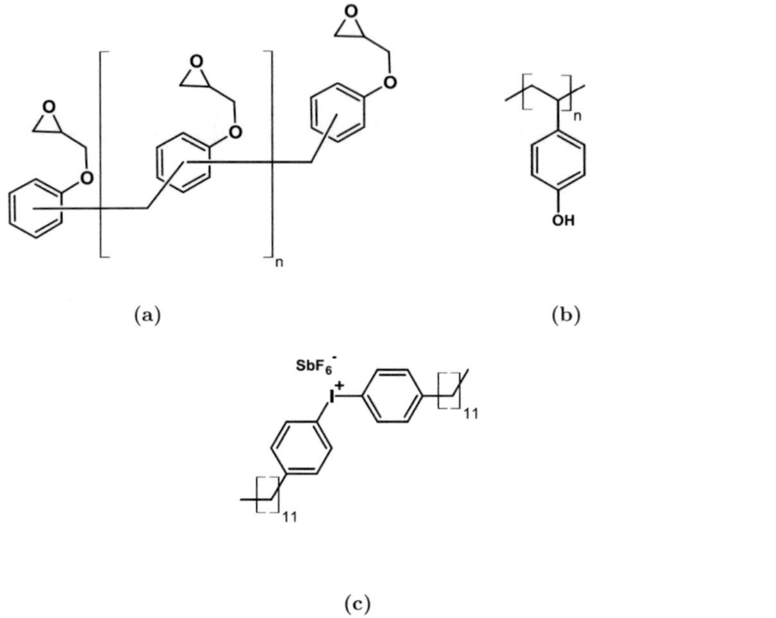

Figure 7.1: Structure of the raw materials used in the cationic formulations: (a) poly[(phenyl glycidyl ether)-co-formaldehyde] (DEN.425) (b) poly(4-vinylphenol) (PVP) (c) bis(4-dodecylphenyl)iodonium hexafluoroantimonate (PAG)

7.1.2 Curing process

The cationic crosslinking process of DEN.425 and PVP was performed in two steps: First, the formulations were exposed to UV-irradiation. In this step, the PAG is activated and a super acid ($HSbF_6$) is produced. Then, the films were further cured on a hot plate. The photo-curing of cationic formulations was only possible on hydrophobic substrates e.g. P3HT. On glass substrates, the adsorbed humidity on the surface was sufficient to inhibit the reaction. The films were not cured when the exposure to UV-irradiation was longer than 30 s. Since the cross-linking reaction needs thermal energy, this reaction does not occur during the exposure to UV-irradiation. In this period of time, other parasitic reactions inhibit the curing reaction by the consumption of the acid. Films with DEN.425 as monomer blended with PVP at different concentrations, PAG (3 wt.%) and Efka 3600 (1 wt.%) was

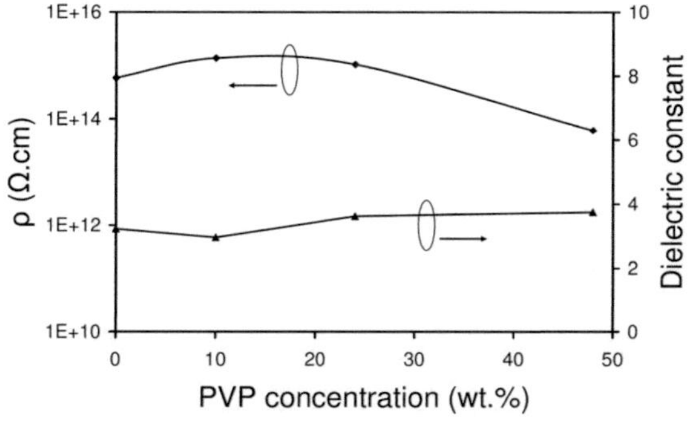

Figure 7.2: Volume resistivity and dielectric constant of films made with DEN.425 blended with different concentrations of PVP [15].

spin-coated on glass substrates coated with Cr/Au. These films were cured for 10 s under the UV-lamp and than at 90 °C for 5 min.

As shown in Fig. 7.2 films made with DEN.425 have high volume resistivity (10^{14} - 10^{15} Ω *cm*). The addition of PVP improve the wetting of the formulations on the substrates and therefore the quality of the films. However, the volume resistivity decreased for high concentrations of PVP. The relative dielectric constant of the films was between 3 and 4. All the films were insoluble in common solvents, e. g. acetone and toluene.

7.1.3 CV-measurement

Top-gate MIS structures made with the cationically curable blends were fabricated and characterized with CV-measurements. In the range between -40 V and +40 V a constant capacitance was measured. This means that the depletion of the semiconductor does not occur in that range. The same result was measured for different concentrations of PAG (1-3 wt.%) and PVP (0-50 wt.%). In the other hand, CV-characteristics of bottom-gate MIS-structures showed a flat-band voltage of 15-16 V. The difference between the results measured in the two structures is that in the bottom-gate structure, the semiconductor gets the contact only with the surface of the cured dielectric material, however, in the top-gate structure the semiconductor get contact with all the chemicals including the super acid in the liquid form. Superacids produced by photo-acid generators were reported to be good dopants for organic semiconductors [103], and therefore they can dope P3HT at the interface. It is also noticeable that the doping concentration (in the bulk) calculated from the CV-measurement of the bottom-gate MIS-structure was equal to 1.8 $\times 10^{16} cm^{-3}$. This value is consistent with the values measured with dielectrics based on P2VP and silres cured with DBU.

DEN.425 was used already in other formulations (see section 6.1), and it did not caused any shift of V_{fb}. It means that this shift is pricipally caused by the PAG. And therefore, cationically photo-cured epoxy could not be used for high performance top-gate transistors.

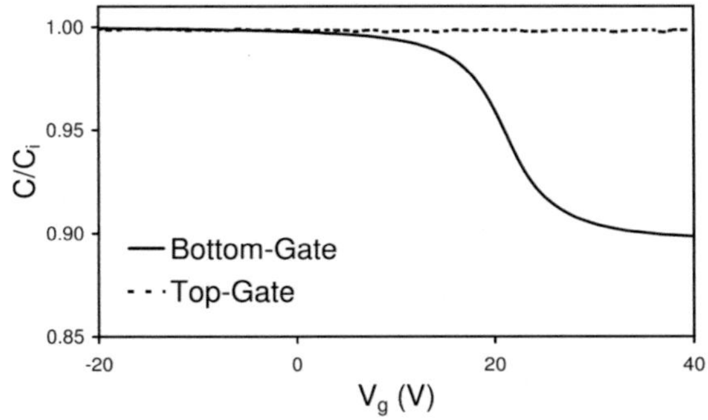

Figure 7.3: CV characteristics of MIS structures made with cationically cured films and P3HT [15].

7.2 Photocurable dielectrics based on acrylates

This section is based on a former own publication [104], therefore data, figures and paragraphs are adopted from it.

7.2.1 Raw materials

As already discussed in section 4.2.2, formulations used for photocurable dielectric should contain monomers, photo-initiators and additives.

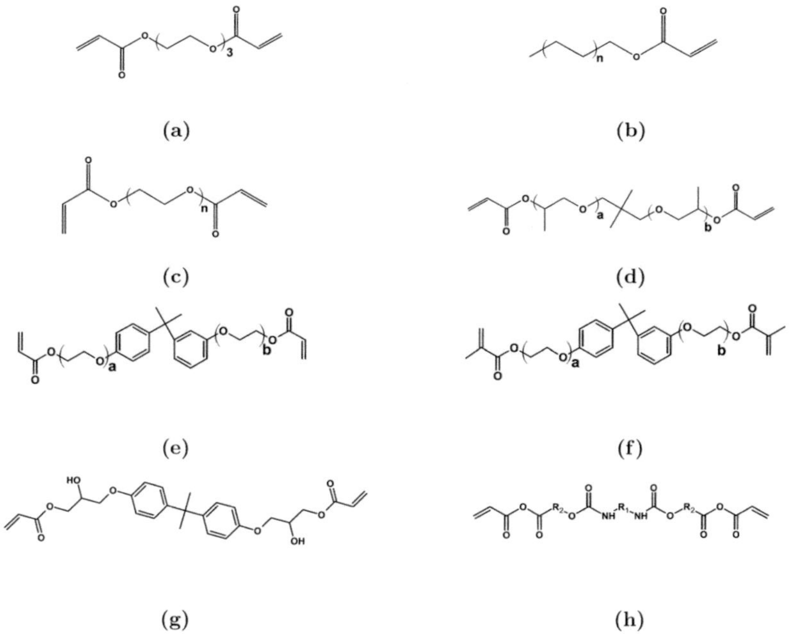

Figure 7.4: (a) 1,6-Hexanediol diacrylate (b) C16, C18 alkyl acrylate (SR257C) (c) Polyethylene glycol diacrylate (d) Propoxylated neopentyl glycol diacrylate, $a + b = 2$ (SR9003) (e) Ethoxylated bisphenol A diacrylate, $a + b = 3$ (SR349) (f) Ethoxylated bisphenol A dimethacrylate, $a + b = 3$ (SR348C) (g) Bisphenol A diglycidylether diacrylate (BDGDA) (h) Desmolux 2513, $R_i = C_nH_{2n}$

Monomers and oligomers

The monomers investigated in this work were chosen upon their properties like number of functionality, viscosity, surface tension, purity and reactivity. As a compromise between reactivity, grade of cross-linking and shrinkage, only bifunctional monomers were used. Another important criteria is to have a low surface tension to be able to wet the semiconductor (P3HT).

1,6-Hexanediol diacrylate (HDDA) was purchased from Alfa Aesar (97%) (Fig. 7.4 (a)).This monomers has a very low viscosity (7 mPa.s). Therefore it is widely used as reactive diluent, which means it is used to reduce the viscosity of formulations and it reacts with the other components during the cross-linking. The surface tension of HDDA is 35.7 mN/m (according to Sartomer Application Bulletin).

C16,C18 alkyl acrylate (SR257C) was supplied by Sartomer (Fig. 7.4 (b)). It could be used in formulations to reduce the surface tension. Because of its monofunctionality, SR257C is used only as additive in this work.

Polyethylene glycol diacrylate (Fig. 7.4 (c)) with two different molecular weights: PEG200DA (n=4) and PEG600DA (n=13-14) were supplied by Double Bond Chemical. PEG200DA is water dispersible and has low viscosity (25 mPa.s). PEG600DA have higher viscosity (80 mPa.s) and is soluble in water. Both monomers, according to the producer, produce soft and flexible films.

Propoxylated neopentyl glycol diacrylate (SR9003) (Fig. 7.4 (d)) is supplied by Sartomer. It is an irritant free monomer with excellent wetting properties because of its low surface tension (32 mN/m). It has a low viscosity (15 mPa.s) and produce flexible films.

Ethoxylated bisphenol A di(meth)acrylates (SR348C and SR349) are supplied by Sartomer (Fig. 7.4 (e, f)). They have a very low volatility and a high viscosity: 1100 mPa.s and 1500 mPa.s respectively. They produce films with good hardness because of aromatic structure. SR349 is also irritant free.

Bisphenol A diglycidylether diacrylate (Genomer 2263) is supplied by Rahn (Fig. 7.4 (g)). It has a high reactivity and provides good hardness properties. It is also an irritant free monomer with very low volatility and high viscosity (3000-6000 mPa.s).

Unsaturated aliphatic urethane acrylate (Desmolux XP 2513) is supplied by Bayer (Fig. 7.4 (h)). It has high reactivity and high viscosity (25000 mPa.s).

PolyQ 9300 is supplied by Double Bond Chemical. It is a pre-polymer which could be cured under UV without use of photo-initiators. It is the only monomer used in this work with a disclosed structure. It has a high viscosity (600-2400 mPa.s) and high reactivity.

Photo-initiators

4-Phenylbenzophenone (PBZ) were supplied by Rahn GmbH, N-methyldiethanolamine (MDEA) and 2-Benzyl-2-dimethylamino-1-(4-morpholinophenyl)-butanone-1 (BDMM) was purchased from Sigma-Aldrich. PBZ is used as Norrish Type II initiator and MDEA as one efficient hydrogen donor and as an oxygen scavenger. BDMM is used as a Norrish Type I initiator.

Additives

Polyether diamine (PEDA) and Doublemer 9136 were supplied by BASF and by Double Bond Chemical respectively. PEDA was used first to control the acid number but it revealed out that it enhance the wetting properties of the formulations. Doublemer 9136 is one wetting agent which contains an acrylate group. Therefore it enhances the substrate wetting and also becomes part of the cross-linked system.

7.2.2 Curing process

Monomer	Composition	ρ [Ω cm]	ϵ_r
PolyQ 9300	100:0:0	2.4×10^{15}	3.8
Genomer 2263	100:2:2	1.2×10^{15}	3.9
SR 349	100:3:3	4.1×10^{14}	5.1
Desmolux 2513	100:3:3	1.2×10^{14}	3.85

Table 7.1: Composition (Monomer : PBZ : MDEA) and dielectric properties of the acrylic photo-cured polymers.

PolyQ 9300 was used without additives, the other monomers were mixed with a photo-initiator. It was not possible to get tack-free films when BDMM was used as photo-initiator. The combination of PBZ and MDEA was efficient in curing thin films of acrylates in air. N-butylacetate was used to reduce the viscosity when necessary. The blends were coated on glass using a cutter to get thin films and exposed to UV-C irradiation using the EPROM eraser "Dienies LG 18" (16 mW/cm^2). The goal is to get hard films after exposure to UV for maximum 2 minutes and using maximum 5 wt.% each of PBZ and MDEA. None of the monomers with low viscosity (1,6-hexanediol diacrylate, PEG200DA and SR9003) could be cured under these conditions. The oxygen inhibition can be more pronounciated in films with low viscosity since the the diffusion of the atmospheric oxygen coud be faster than its consumption through radicals [105]. The monomer containing a methacrylate group (SR348C) could not also be cured in the same conditions. It was already shown that methacrylates have lower reactivity than the acrylates [106]. The other formulations (see table 7.1) were spin-coated on double-side polished silicon wafers to get films with a thickness of approximatly 1μm. The curing process of them was monitored using FT-IR spectroscopy.

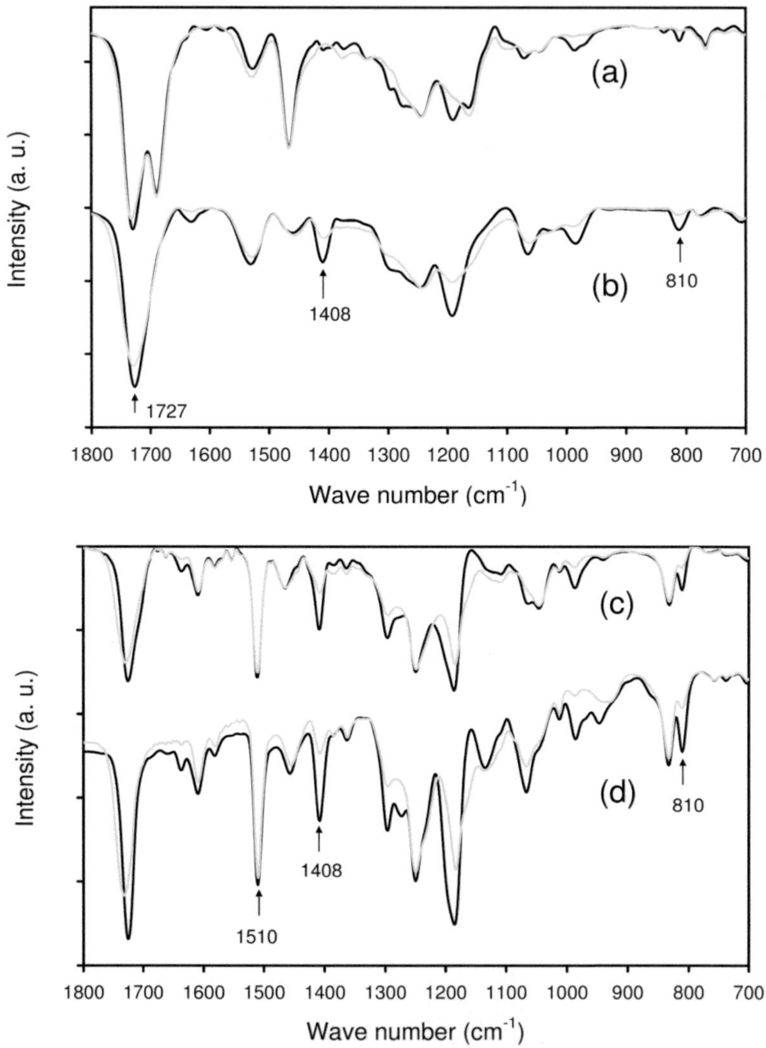

Figure 7.5: FT-IR spectra of (a) Desmolux 2513, (b) PolyQ 9300, (c) SR349 and (d) Genomer2263 before (black) and after (grey) 60 s of exposure to UV-Irradiation.

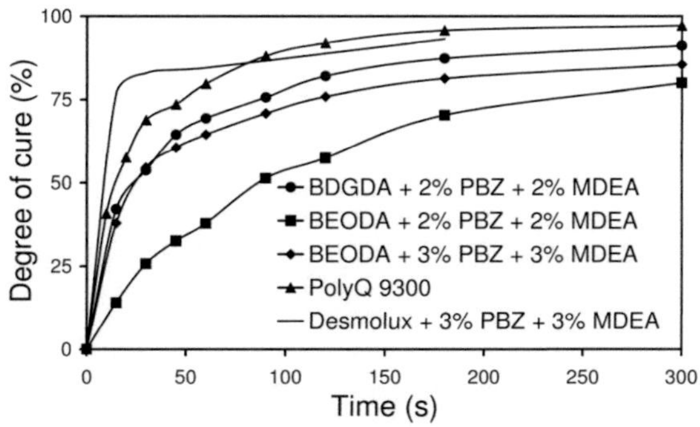

Figure 7.6: Degree of conversion versus time of the acrylate group in different formulations upon exposure to UV-irradiation in air.

The conversion of the acrylate group in the formulations was calculated by measuring the rate of disappearance of the $= CH_2$ twisting mode at 810 cm^{-1} or the deformation mode of $= CH_2$ at 1408 cm^{-1} [100]. The absorption peak of the aromatic ring at 1510 cm^{-1} and the absorption peak of the carbonyl group at 1727 cm^{-1} [100] were used as references for the calculation of the relative disappearance of the acrylate group for respectively the monomers containing Bisphenol A in their structures and the two other monomers. Fig. 7.6 shows the degree of conversion of the acrylate group in the formulations based on the four monomers upon exposure to UV-irradiation. In the case of Desmolux 2513 no reliable data could be measured for films with $1\mu m$ thickness since the liquid does not wet the substrate even after adding the wetting agent. The film is transformed into many drops on the silicon wafer. Due to their higher thickness, these drops cured faster than the other formulations. On glass, as in the case of SR349, 3 wt.% of initiators was necessary to get tack-free films within one minute of irradiation.

MIM-structures were fabricated using the formulations exhibiting at least 70 % of converted acrylate groups after one minute of curing. According to Fig. 7.7, all the fabricated films have leakage currents lower than 10^{-8} $A\ cm^{-2}$ at 1 $\frac{MV}{cm}$ corresponding to a minimum volume resistivity of 10^{14} $\Omega\ cm$. PolyQ 9300 and Genomer 2263 have the highest volume resistivity of 2.4×10^{15} $\Omega\ cm$ and 1.2×10^{15} $\Omega\ cm$ respectively. These values are higher then the ones measured for linear polymers as shown in chapter 5. As a consequence of these results these materials could be considered as candidates to be used as dielectrics for organic transistors.

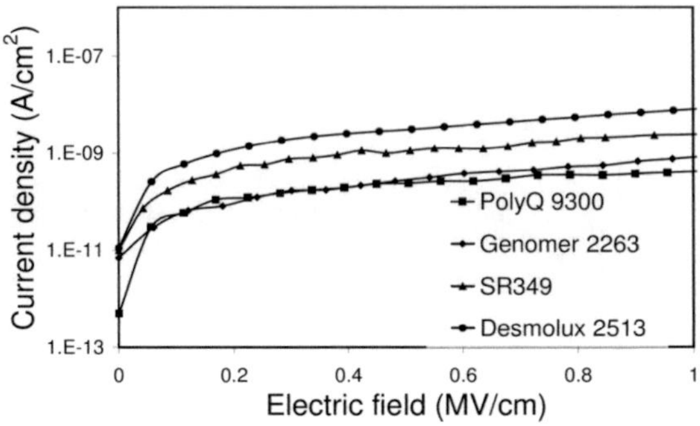

Figure 7.7: Leakage current density versus applied voltage for films made with blends containing PolyQ 9300, Genomer 2263, SR349 and Desmolux 2513.

7.2.3 CV-measurement

As top gate structures are studied in this work, the semiconductor is exposed to the UV-irradiation and to the reactive radicals during the cross-linking reaction. It is important to study their effect on the dielectric/semiconductor interface. It is expected that the UV-irradiation would fill and maybe generate trap states at the semiconductor / insulator interface, since there is many reports about the effect of UV treatment on organic transistors or MIS-capacitors [107, 108].

The first measurement with PolyQ 9300 confirmed this assumption. Fig. 7.8a showed a CV-characteristic of one MIS-structure made with PolyQ 9300 which was cured under UV-irradiation for 1 min. The CV-characteristic exhibits one large positive shift of the V_{fb}. This could be correlated with high density of trap states at the interface between semiconductor and the insulator. The fact that this shift is still measurable even many hours (time needed for the evaporation of the upper electrode) after the irradiation means that these trap states have long life time.

The interface between the semiconductor and the insulator could be engineered when wetting agent are added to the formulation since the wetting agent would adsorb at semiconductor interface [109]. In fact, the wetting agent is also necessary to get thin films of acrylates on the P3HT in a transistor because of the very low surface tension of P3HT. The addition of low concentration of polyether diamine (PEDA) enhanced the wetting and the gloss of PolyQ 9300 on P3HT. The quality of the interface between P3HT and the mixture PolyQ 9300 with PEDA was investigated by studying the effect of PEDA on the V_{fb} of MIS-structure made with PolyQ 9300. Two different trends could be seen in Fig. 7.8b, the first one is that PEDA reduces the V_{fb} and that prolonging the time of exposure to UV-irradiation increases the V_{fb}. The dependence of the V_{fb} on the time of exposure is almost negligible in the case of the pure PolyQ 9300. The fact that other neutral wetting agent like the Doublemer 9136, which is an acrylate functionalized silicone, reduces also the V_{fb}, excludes the statement that the effect of PEDA on the V_{fb} is caused by its basicity. In the coming experiments, Doublemer 9136 is used as wetting agent (1 *wt.*%) because it can be incorporated in the matrix of the cross-linked system due to the acrylate functionality.

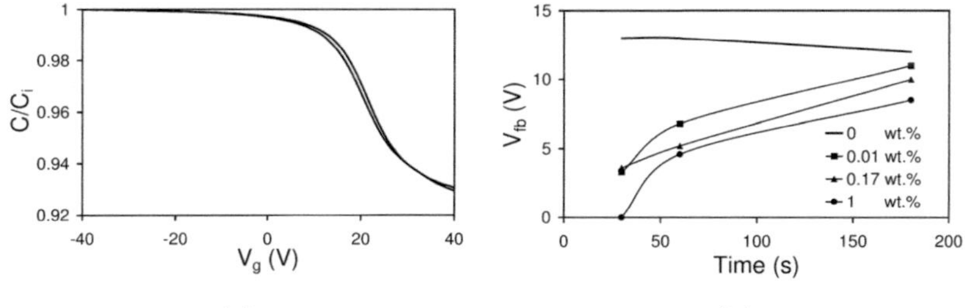

Figure 7.8: (a) CV-characteristic of a MIS-structure made with PolyQ 9300 and cured for 1 min under UV-irradiation. (b) CV-characteristic of a MIS-structure made with PolyQ 9300 blended with different concentrations of PEDA and cured with different time of exposure to UV-irradiation.

The effect of doublemer 9136 and of annealing on the CV-characteristics of MIS-structures made with PolyQ 9300 and Genomer 2263 was investigated as shown in the Fig. 7.9. In the case of PolyQ 9300, annealing the structures at 90 °C was sufficient to reduce the V_{fb}. The addition of doublemer 9136 to the resin before the UV-irradiation could also reduce the V_{fb} to 2.6 V. It is noticeable that all the MIS-structures made with PolyQ 9300 showed no hysteresis even in air. On the other hand, Genomer 2263 caused in air a hysteresis with a width of $\Delta V_g = 41$ V. In vacuum or in nitrogen, a negligible hysteresis could be measured. The measured hysteresis could be caused by the water absorbed by the polymer. The OH group contained in the Genomer 2263 could be responsible for the hygroscopicity of the films. In vacuum, the structure still have high V_{fb} only after annealing at 90 °C, the V_{fb} is reduced. The addition of Doublemer 9136 to Genomer 2263 have two

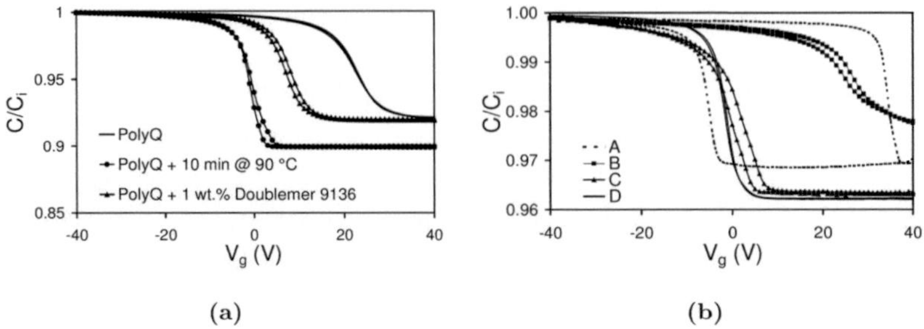

Figure 7.9: (a) CV-characteristics of MIS-structures made with PolyQ 9300 before and after annealing for 10 min at 90 °C. All the measurements are made in air. (b) CV-characteristics of MIS-structures made with Genomer 2263 in air (A), in vacuum (B), after annealing for 20 min at 90 °C in vacuum (C) and mixed with 1 wt.% of 9136 in air (D).

effects: the disappearance of the hysteresis even in the air and the reduction of the V_{fb} to -2.6 V even without annealing. The disappearance of the hysteresis could be correlated with the increase of the hydrophobicity of the films, since the non-polar group of the wetting agent would be oriented toward the air. The reduction of the shift of V_{fb} means a better interface between the dielectric and the semiconductor when the wetting agent is used. The doping concentration calculated from the $1/C^2$-V slope for the MIS-structures containing Doublemer 9136 was equal to around 2 $\times 10^{16}$ cm^{-3}, which is similar to the value calculated from CV-curves of MIS-capacitors made with P2VPy and P2VPy-DEN.425 mixtures.

7.2.4 Transistors

The transfer characteristics of the transistors made with acrylates (Fig. 7.10 (b)) show high on/off ratios (10^5 - 10^6). That is due to the low off-currents of all the transistors especially in the case of Genomer 2263. At $V_{gs} = 0$ V the off-current is around 10^{-12} A for transistors made with Genomer 2263. In the case of the other monomers all the transistors have off-currents between 10^{-11} and 10^{-10} A. In order to calculate the field-effect mobilities and the threshold voltages of the devices, the transfer characteristics was fit using the equation 2.6. The field-effect mobilities of P3HT at $V_{gs} = -40$ V was approximately equal to 10^{-3} cm^2 V^{-1} s^{-1} wenn PolyQ 9300, Genomer 2263 and SR 349 were used. The mobility was nine times higher in

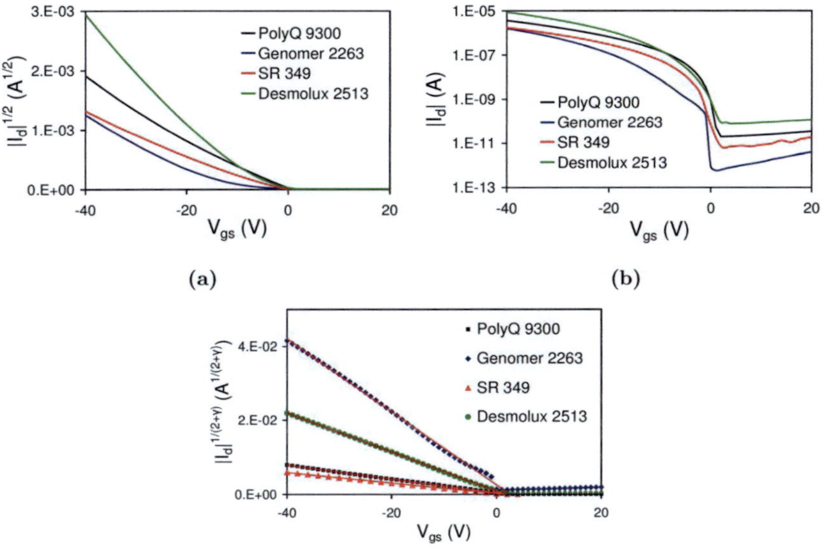

Figure 7.10: (a), (b)Root squared and logarithmic transfer characteristics of transistors made with PoyQ 9300, Genomer 2263, SR 349 and Desmolux 2513. (c)Plot of measured current raised to the power $1/(2+\gamma)$.

the case of Desmolux 2513, which explains the high on-current of the corresponding transistor. In the other hand, the threshold voltages of the transistors have almost the same value. This result could be correlated with the measurement of MIS-structures, which shows that the structures have also almost the same V_{fb}. It is also noticeable that the transistors made with PolyQ 9300 and Genomer 2263 show low sub-threshold swings (0.85 V/dec and 0.44 V/dec respectively), which means that they have low density of interface traps. According the equation:

$$N_{SS} = \left[\frac{S\ log\,(e)}{\frac{\kappa\ T}{q}} - 1 \right] \frac{C_i}{q} \tag{7.1}$$

both transistors have trap density in the order of $10^{11}\ cm^{-2}$. All the transistors have much higher on/off ratios than the polystyrenic-based transistors. However, only transistors based on Desmolux 2513 have field-effect mobilities in the same range of the polystyrenic-based transistors. Table 7.2 summarizes the electrical data of the transistors made with the acrylic formulations.

Monomer	ϵ_r	γ	μ @ -40 V $[cm^2\ V^{-1}\ s^{-1}]$	On/off	SS $[V/dec]$	N_{SS} $[cm^{-2}]$	V_{th} $[V]$
PolyQ 9300	3.8	0.59	1.9×10^{-3}	1.9×10^{5}	0.85	6.9×10^{11}	-2.24
Genomer 2263	3.9	2.2	1.3×10^{-3}	2.6×10^{6}	0.44	3.7×10^{11}	-2.5
SR 349	5.1	0.59	1.1×10^{-3}	2.4×10^{5}	1.42	1.1×10^{12}	-0.8
Desmolux 2513	3.85	1.05	9.0×10^{-3}	1.1×10^{5}	2.4	1.5×10^{12}	-1.47

Table 7.2: Electrical properties of top-gate transistors fabricated with UV-curable acrylates as dielectric materials.

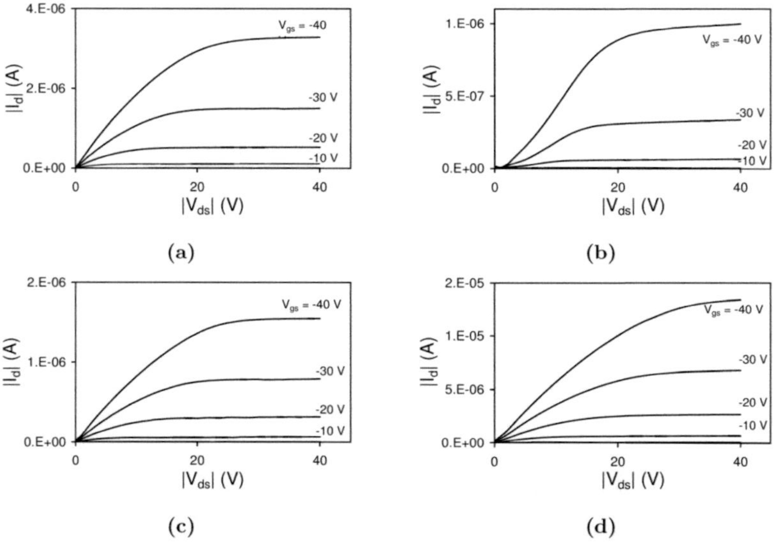

Figure 7.11: Output characteristics of transistors made with (a) PoyQ 9300 (b) Genomer 2263 (c) SR 349 (d) Desmolux 2513.

Fig 7.11 shows the output characteristics of the transistors fabricated with acrylic insulators. All the transistors exhibited a good saturation behavior. The most important difference between the output characteristics of the different transistors, is that the output characteristic of the transistor made with Genomer 2263 exhibits a nonlinearity near $V_{ds} = 0\ V$. This could be correlated with the high value of γ which indicates the high dependance of the field-effect mobility on the electrical field. Scheinert *et al.* showed that the combination of Schottky-barriers at the source/drain contacts and the dependance of the mobility on the electrical field and the density of carrier can lead to a nonlinearity in the output characteristics [110]. The transistors based on Desmolux 2513 exhibit a slightly nonlinearity for low V_{ds}. On the other hand, the output characteristics of transistors having low γ values are linear at low V_{ds}.

7.3 Thiol-ene based dielectrics

The third approach to produce photo-curable dielectrics for top-gate OFETs is based on thiol-ene systems. This section is based on a former own publication [111], therefore data, figures and paragraphs are adopted from it.

7.3.1 Raw materials

Monomers

Multifunctional monomers, from which at least one is tri-functional or more, are needed to obtain cross-linked thiol-ene polymers. PETMP with low concentration of 3-mercaptopropionic acid (< 0.1 %) was supplied by Bruno Bock GmbH and used without purification. The concentration of the non-reacted acid was approximated from the acid number of the monomers provided by the producer (acid number = 0.56 mg KOH/g).
Polysiloxanes with pending mercapto groups could be also used for photo-initiated thiol-ene polymerizations [112, 113]. These hybrid polymers are synthesized with mercapto-functionalized dialkoxysilanes (or trialkooxysilanes) as precursor with sol-gel processes. Alkoxysilanes are hydrolysed in acidic or basic conditions (see Fig. 7.12 a). The produced silanol can undertake a water or an alcohol producing condensation reaction (see Fig. 7.12 b-c). Poly(3-mercaptopropyl methylsiloxane) (PMMS) was prepared by sol-gel process and used in this work as multifunctional thiol. (3-Mercaptopropyl)methyldimethoxysilane (MDMS, 95 %) was purchased from Alfa Aesar. Hydrochloric acid (32 % aqueous) was purchased from Merck, poly (4-vinylpyridine) (MW: 200,000) from Scientific Polymer Products and 2-propanol from Carl-roth. MDMS (5 g, 0.038 mol) was mixed with water (1.36 g, 0.076 mol), HCl (0.0038 mol) and 5 g of 2-propanol under magnetic stirring at room temperature for 1 h. Then the mixture was heated at 80 °C for 3 h. Poly(4-vinylpyridine) was added

in excess to scavenge the acid. The mixture was filtered through 0.2 m PTFE-filter and then heated to 100 °C till the solvents (water and alcohols) were removed to afford a viscous liquid.

There is a wide range of possible ene for the photo-polymerized thiol-enes, since any kind of ene including norbornene, vinyl-ether, acrylate and alkene could be used [97]. Cramer *et al.* reported that vinyl-ether group is very reactive in thiol-ene systems [114]. 1,4-Cyclohexanedimethanol divinyl ether (CHDM-DVE), a small monomer with a high boiling point (253 °C), was used in this work as an ene in the thiol-ene formulations. CHDM-DVE (97 %) was supplied by BASF and used without further purification. Vinyl-functionalized polysiloxane (PVS) was also produced by sol-gel with methydimethoxyvinylsilane (97 %) in the same way used for the synthesis of PMMS, as shown in Fig. 7.12.

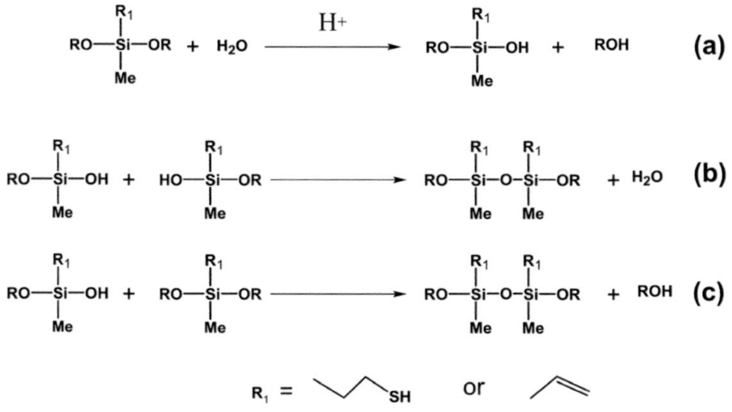

Figure 7.12: Schematics of the synthesis of PMMS and PVS.

Additives

Irgacure 127 is a double functional alpha-hydroxyketone, this class of photo-initiators has good surface curing properties [115]. Therefore it is used here to initiate the thiol-ene reaction. PMMS and PVS could wet P3HT without any additive. However, PETMP has a higher surface tension and thus Doublemer 9136 was mixed with PETMP to get thin films on the top of P3HT. Fig. 7.13 shows the structure of the monomers and the photo-initiator used in this work.

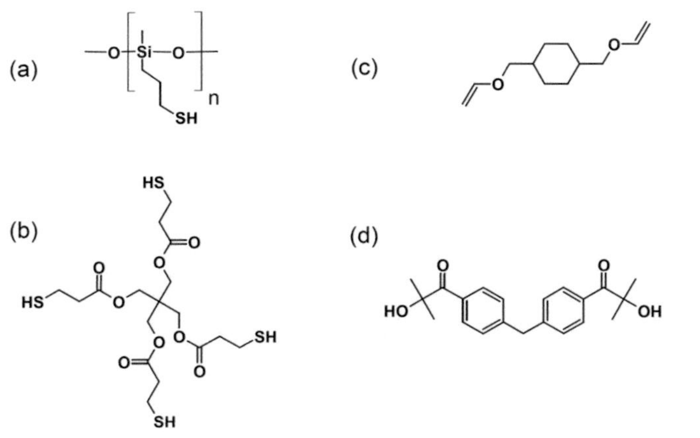

Figure 7.13: Structure of (a) PMMS, (b) Pentaerythritoltetra (3-mercaptopropionat) (PETMP), (c) 1,4-Cyclohexanedimethanol divinyl ether (CHDM-DVE) and (d) irgacure 127.

7.3.2 Curing process

Three thiol-ene systems were investigated: PMMS with CHDM-DVE, PMMS with PVS and PETMP with CHDM-DVE. The formulations was blended with Irgacure 127 (5 $wt.\%$) and the photo reaction is initiated by UV-C irradiation (254 nm).
FT-IR spectra of films made with the blends before and after exposure to UV-irradiation are shown in Fig.7.14. To calculate the degree of cure of the films, the disappearance of the doublet peak of the vinyl ether in the region 1640-1610 cm^{-1} was monitored [100]. The strong absorption peak of the siloxane -$(SiO)_n$- around 1070 cm^{-1} and the peak of the carbonyl group around 1735 cm^{-1} were used as references [100]. Fig. 7.14 shows that the reaction of the vinyl-ether with the thiol is fast. After 5 s of UV-irradiation more than 85 % of the vinyls in CHDM-DVE reacted with PMMS. Further 10 s were needed to get tack-free films, the degree of cure was about 89 %. Further irradiation did not increase this value. Films made with PMMS and CHDM-DVE irradiated for 15 s reached an excellent resistivity value of 2×10^{15} Ω cm.

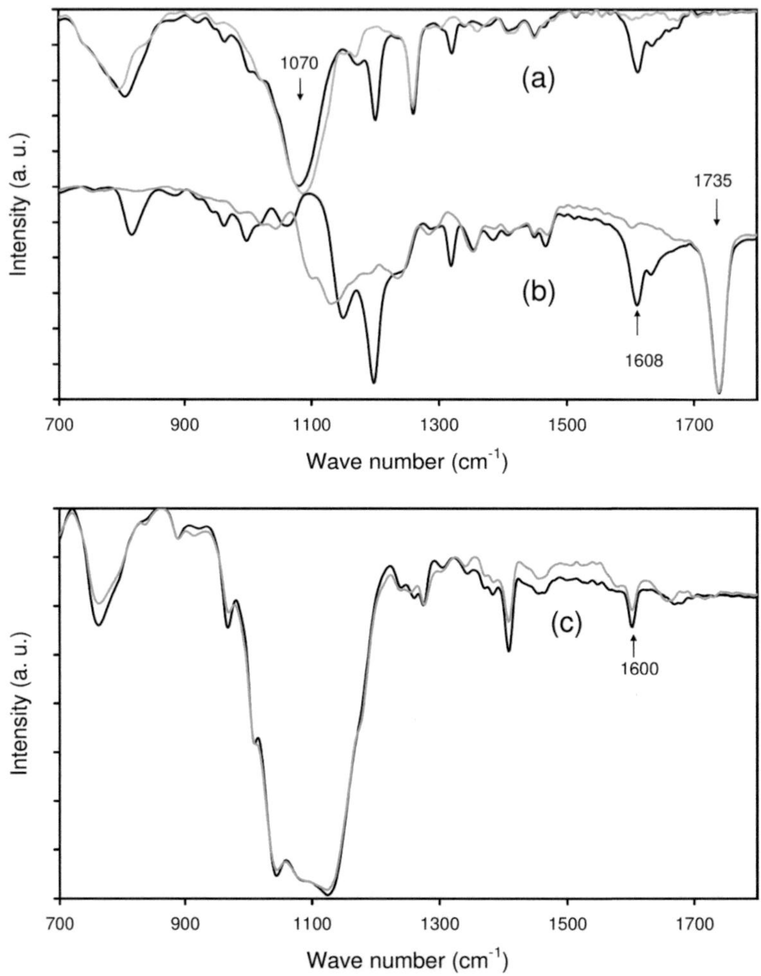

Figure 7.14: FT-IR spectra of (a) PMMS with CHDM-DVE, (b) PETMP with CHDM-DVE, (c) PMMS with PVS before (black) and after (grey) exposure to UV-Irradiation (5s for (a) and (b); and 60 s for (c)).

Longer exposure did not improve this value since at 15 s the maximum of the degree of curing of the vinyl-ether is already reached. On the other hand, PETMP showed a degree of cure higher than 97 % after 5 s of irradiation. Films made with

PMMS with PVS could not be cured even after many minutes of irradiation. It was not possible to get tack-free films made with PMMS-PVS mixtures.

7.3.3 CV-measurement

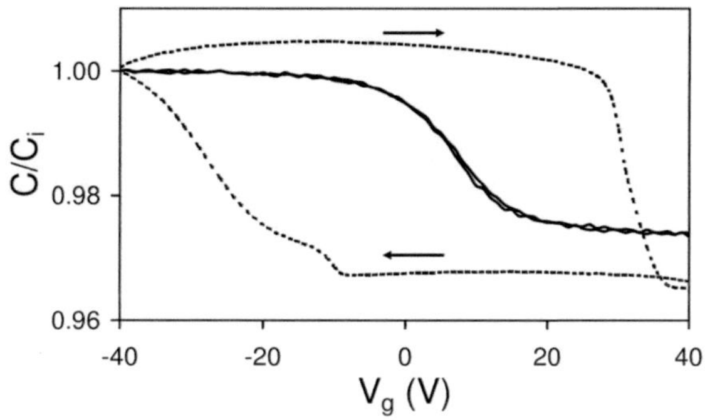

Figure 7.15: CV curves (1 kHz) measured on Au/P3HT/polymer/Au MIS-structures using thiol-ene mixtures made with PMMS (continuous) and PETMP (dashed).

As the semiconductor is exposed to the UV irradiation and to the reactive radicals during the cross-linking reaction, it is important to study their effect on the dielectric-semiconductor interface. For this purpose, CV characteristics of MIS structures containing thiol-ene mixtures as dielectric material were measured (Fig. 7.15). The mixture made with PETMP produce one large higher back sweep current (higher BSC) hysteresis. The higher BSC hysteresis could be caused by mobile ions or by the polarization of the dielectric [116]. The PETMP still contain a high concentration of the base used for neutralization (57 mg/g NH_4^+ measured with steam distillation and titration using a Büchi B323 N2-Analyzer [1]), this high concentration of ions could be responsible for the higher BSC hysteresis shown in Fig. 7.15 [117]. Egginger *et al.* showed that the presence of ions in poly(vinylalcohol) used a insulator in a transistor caused a high BSC hysteresis in its transfer characteristics. The MIS structure made with PMMS showed no hysteresis, this is due to the low density of the mobile ions in the dielectric. The apparent doping concentration calculated from the $1/C^2$-V slope for the MIS-structures containing PMMS was equal to around

[1]Measurement performed by the central laboratory of the Hamburg University of Technology.

$7 \times 10^{16}\ cm^{-3}$, this value is about three times higher than the value measured for MIS-structures made with P2VPy. Since the same batch of P3HT was used in both cases, it is assumed that the thiol-ene polymers causes more interface traps on P3HT than P2VPy. The V_{fb} was estimated to be equal to -1 V. As consequence of these results, only the mixture made with PMMS was further characterized.

7.3.4 Transistors

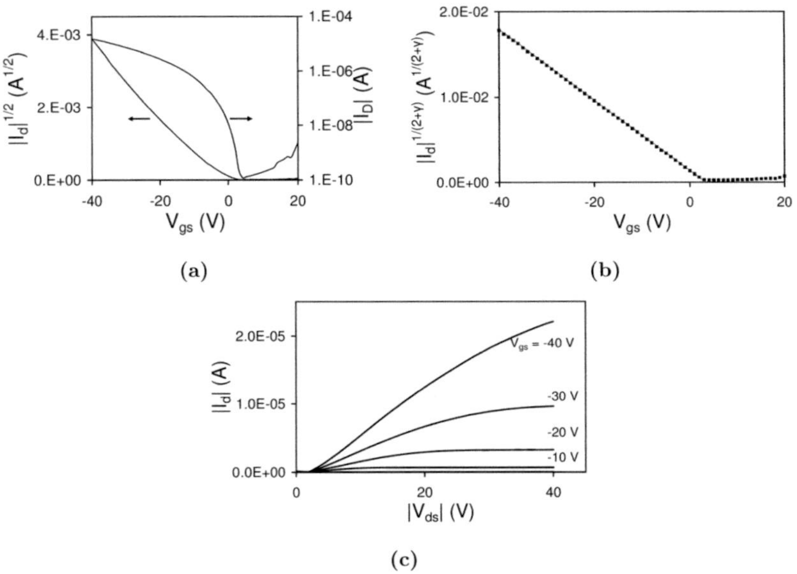

Figure 7.16: (a) Root squared and logarithmic transfer characteristics of transistors made with PMMS and CHDM-DVE, (b)plot of measured current raised to the power $1/(2+\gamma)$, (c) output characteristics of the transistor.

Top-gate OFETs were fabricated using a mixture of PMMS, CHDM-VDE and irgacure 127. The dielectric material was exposed for 15 s to UV-irradiation. The transfer characteristic of the transistor (Fig. 7.16 a) shows a high on/off ratio of more than 10^5. The sub-threshold swing was equal to 1.25 V/dec, corresponding to a traps density of about $10^{12}\ cm^{-2}$. The field effect mobility μ and the threshold voltage V_{th} were calculated by fitting the plot of measured current raised to the power $1/(2+\gamma)$ versus the gate voltage (Fig. 7.16 (b)). The dielectric constant used

Monomer	ϵ_r	γ	μ @ -40 V $[cm^2\ V^{-1}\ s^{-1}]$	On/off	SS $[V/dec]$	N_{SS} $[cm^{-2}]$	V_{th} $[V]$
PMMS + CHDM-DVE	4.6	0.75	8.8×10^{-3}	1.25×10^5	1.25	1×10^{12}	3.1

Table 7.3: Electrical properties of top-gate transistors fabricated with thiol-ene polymer as dielectric material.

in the calculation was determined from the MIM structure and was equal to 4.6. The field effect mobility and the threshold voltage determined from the saturation region were 8.8×10^{-3} $cm^2V^{-1}s^{-1}$ at $V_{ds} = V_{gs} = -40\ V$ and 3.1 V respectively. Table 7.3 summarizes the electrical data of the transistors made with the thiol-ene formulation. The transistors based on the thiol-ene polymers showed better on/off ratios and sub-threshold swings than the P4tBS-based transistors. However, their field-effect mobilities was slightly lower than those of the polystyrenics-based transistors. Unlike transistors made with acrylates, the nonlinearity in the output characteristics (Fig. 7.16 c) could not be correlated with the value of γ. These results show that cross-linked polymers based on thiol-ene can be used to produce high performance transistors despite their high reactivity.

7.4 Conclusion

In this chapter, three different approaches of new photo-curable dielectric materials for top-gate transistors were presented. The first one was to use a monomer with epoxy-functionality blended with a PAG. The presented formulation cured fast under UV-irradiation and heat treatment. It revealed out that the PAG produce a high shift of V_{fb} which makes this approach not useful for high performance transistors. The second one was to use free radical photo-curable dielectrics based on acrylates. Although these formulations was cured in the air for only one minute, they produced films with good dielectric properties. The transistors based an acrylates had high on/off ratios and low sub-threshold swings. The reaction of the polymerization upon the semiconductor did not produce any permanent shift of V_{fb}. The third one was the use of thiol-ene polymers. These polymers could cure much faster than all the other formulations used in this work. Films made with thiol-ene formulations could be cured even within 5 seconds. A mercapto-functionalized polysiloxane mixed with a vinylether and a photo-initiator was used a dielectric material. This new insulator showed high volume resistivity and high performance in top-gate transistors. Like in the case of the thermally cured polymers, transistors based on the photo-cured polymers have much higher on/off ratios than the transistors based on the uncross-linked polymers. The mobilities of only Desmolux 2513 and the thiol-ene polymer based transistors are in the same range with the field-effect mobilities of the polystyrenic-based transistors.

8. Summary and outlook

The development of fast curing cross-linked polymers could further simplify the fabrication processes of organic circuits. In this work, the curing processes and the electrical properties in capacitors and transistors of solution-processed dielectrics were investigated. The main goal was to reduce the time and steps of the curing process to enable fast printing of circuits for high throughput production without loss of performance. In chapter 6, two approaches of thermally cured polymers were presented: Poly(2-vinylpyridine) (P2VPy) cross-linked with multi-functional epoxides and a methoxy-substituted polysiloxane cross-linked with acids or bases. The cross-linking reaction of the P2VPy requires the thermal treatment for 30 min at 140 °C. The polysiloxane was cured at least after one hour at 150 °C if bases or low concentration of an acid were used as catalysts. The curing time could be reduced by increasing the concentration of the acid. The CV-characterization of MIS-capacitors made with these materials showed that MIS-capacitors based on uncured P2VPy, cured P2VPy and polysiloxane cured with a base have a flat-band voltage of about -6 V. When an acid is used to cure the polysiloxane, the flat-band voltage was positive and increases with increasing acid concentration. The transistors made with these materials have higher on/off ratios and lower mobilities than transistors made with linear polystyrenic polymers (studied in Chapter 5). These materials could substitute the linear polymers only in applications in which high chemical stability is required and the speed of curing of the formulation is not critical.

In chapter 7, photo-cured polymeric dielectrics were investigated. They were based on monomers which photo-polymerize on the semiconductor when exposed to UV-irradiation. Their polymerization reactions were much faster then the thermally cured ones. Cationic systems were cured after 10 s of UV exposure and thermal treatment of 5 min at 90 °C. However, a high shift of the flat-band voltage in the CV-curve of the corresponding MIS-capacitor was measured. It is assumed that the photo-generated super-acid was responsible for this shift. Therefore, the cationic system is not qualified for use as dielectric for top-gate transistors. The photo-cured polymers based on acrylates could be cured within one minute of exposure to UV-irradiation. The acrylic dielectrics have high volume resistivities and did not produce any shift in the flat-band voltage, if they were mixed with a wetting agent or thermally treated for 30 min at 90 °C. The curing time could be further reduced by using thiol-ene systems which were able to cure within seconds under UV-irradiation. The thiol-ene system was based on a mercapto-substituted polysiloxane prepared with a sol-gel process, a bi-functional vinyl-ether and a photo-initiator. The thiol-ene based dielectric had also high volume resistivity and created a negligible shift of the flat-band voltage. Transistors made with acrylic and thiol-ene based dielectrics have also high on/off ratios (up to 10^6). The field-effect mobilities of holes measured using transistors based on urethane acrylate and the thiol-ene system were in the same range with values measured using polystyrenic-based transistors. Transistors made with urethane acrylate and thiol-ene polymer have slightly lower field-effect mobilities, higher on/off, lower sub-threshold swing and higher chemical stability than transistors based on polystyrenics. Therefore, these two formulations provide the required criteria for the production of high performance transistors.

After the identification of possible reaction mechanisms enabling the fast curing of dielectric materials and their high performance in OFETs, optimization of the structure of the monomers and their purity grade should be done in order to achieve OFETs with higher field effect mobilities. These optimizations could be also done in order to reduce the thickness of the dielectric layers to have low-voltage OFETs. The developed formulations could be also further engineered for encapsulations for top-gate and bottom-gate transistors. It is also important, to custom the formulation properties, e.g. viscosity, for the different high-throughput fabrication processes like ink-jet and roll-to-roll printing.

In conclusion, it is shown that photo-curing reaction involving free radicals is very fast and do not charge or generate traps in the interface between semiconductor and dielectric. Moreover, the use of low viscosity monomers is also beneficial in reducing the use of volatile organic compounds. These properties, and of course the good electrical insulation, make the promising approach of using monomers based on acrylates or on thiol-enes and curing them directly on the semiconductor very advantageous for high-throughput production of high performance organic circuits.

Bibliography

[1] H. Klauk. *Organic Electronics: Materials, Manufacturing, and Applications.* Wiley, 2006.

[2] H. Klauk. *Organic Electronics II: More Materials and Applications.* Wiley, 2012.

[3] F. So. *Organic Electronics: Materials, Processing, Devices and Applications.* CRC Press, 2010.

[4] H. Marien, M. Steyaert, and P.L. Heremans. *Analog Organic Electronics: Building Blocks for Organic Smart Sensor Systems on Foil.* Springer, 2012.

[5] S.S. Sun and L.R. Dalton. *Introduction to Organic Electronic and Optoelectronic Materials and Devices.* Taylor & Francis Group, 2008.

[6] S.S. Sun and N.S. Sariciftci. *Organic Photovoltaics: Mechanisms, Materials, and Devices.* Taylor & Francis, 2010.

[7] C. Brabec, U. Scherf, and V. Dyakonov. *Organic Photovoltaics: Materials, Device Physics, and Manufacturing Technologies.* Wiley, 2011.

[8] T. Komoda, N. Ide, and J. Kido. High efficient OLEDs and their application to lighting. *Journal of Light & Visual Environment*, 32(2):75–78, 2008.

[9] H.D. Kim, H.J. Chung, B.H. Berkeley, and S.S. Kim. Emerging technologies for the commercialization of AMOLED TVs. *Information Display*, 25(9), 2009.

[10] K.J. Kim, E. Park, and S.S. Sundar. IPS vs. AMOLED: Effects of panel type on smartphone users viewing and reading experience. *Human Centric Technology and Service in Smart Space*, pages 77–84, 2012.

[11] www.lg.com.

[12] www.heliatek.com.

[13] Z. Bao and J. Locklin. *Organic Field-Effect Transistors.* Taylor & Francis, 2007.

[14] Data measured by Cenk Akbulak in his study thesis: Untersuchung von Polymer-Dielektrika für organische Feld-Effekt Transistoren.

[15] Data measured by Nils Steller in his diploma thesis: Vernetzte Polymere als Dielektrikum für organische Feld-Effekt Transistoren.

[16] G.W. Neudeck and A.K. Malhotra. An amorphous silicon thin film transistor: Theory and experiment. *Solid State Electron.*, 19(8):721–729, 1976.

[17] A. Tsumura, H. Koezuka, and T. Ando. Macromolecular electronic device: Field-effect transistor with a polythiophene thin film. *Appl. Phys. Lett.*, 49(18):1210–1212, 1986.

[18] S.M. Sze and K.K. Ng. *Physics of Semiconductor Devices.* Wiley, 2006.

[19] D.K. Schroder. *Semiconductor material and device characterization.* John Wiley & Sons, Inc., 1990.

[20] Y. Kuo. *Thin Film Transistor Technologies (TFTT VII): Proceedings of the International Symposium.* Electrochemical Society, 2005.

[21] G. Horowitz, P. Lang, M. Mottaghi, and H. Aubin. Extracting parameters from the current-voltage characteristics of organic field-effect transistors. *Adv. Funct. Mater.*, 14(11):1069–1074, 2004.

[22] M. Vissenberg and M. Matters. Theory of the field-effect mobility in amorphous organic transistors. *Phys. Rev. B*, 57(20):12964, 1998.

[23] C.K. Chiang, C.R. Fincher, Y.W. Park, A.J. Heeger, H. Shirakawa, E.J. Louis, S.C. Gau, and A.G. MacDiarmid. Electrical conductivity in doped polyacetylene. *Phys. Rev. Lett.*, 39:1098–1101, 1977.

[24] A.J. Heeger, N.S. Sariciftci, and E.B. Namdas. *Semiconducting and Metallic Polymers.* Oxford Graduate Texts. OUP Oxford, 2010.

[25] J.L Bredas and G.B. Street. Polarons, bipolarons, and solitons in conducting polymers. *Accounts Chem. Res.*, 18(10):309–315, 1985.

[26] A.K. Bakhshi. Electrically conducting polymers: From fundamental to applied research. *B. Mater. Sci.*, 18:469–495, 1995.

[27] X. Peng, G. Horowitz, D. Fichou, and F. Garnier. All-organic thin-film transistors made of alpha-sexithienyl semiconducting and various polymeric insulating layers. *Appl. Phys. Lett.*, 57(19):2013–2015, 1990.

[28] F. Garnier, R. Hajlaoui, A. Yassar, and P. Srivastava. All-polymer field-effect transistor realized by printing techniques. 265(5179):1684–1686, 1994.

[29] C.J. Drury, C.M.J.Mutsaers, C.M. Hart, M. Matters, and D.M. De Leeuw. Low-cost all-polymer integrated circuits. *Appl. Phys. Lett.*, 73(1):108–110, 1998.

[30] Z. Bao, Y. Feng, A. Dodabalapur, V.R. Raju, and A.J. Lovinger. High-performance plastic transistors fabricated by printing techniques. *Chemistry of Materials*, 9(6):1299–1301, 1997.

[31] C.D. Dimitrakopoulos, B.K. Furman, T. Graham, S. Hegde, and S. Purushothaman. Field-effect transistors comprising molecular beam deposited α, ω-di-hexyl-hexathienylene and polymeric insulator. *Syn. Met.*, 92(1):47 – 52, 1998.

[32] G.H. Gelinck, T.C.T. Geuns, and D.M. de Leeuw. High-performance all-polymer integrated circuits. *Appl. Phys. Lett.*, 77(10):1487–1489, 2000.

[33] M. Halik, H. Klauk, U. Zschieschang, T. Kriem, G. Schmid, W. Radlik, and K. Wussow. Fully patterned all-organic thin film transistors. *Appl. Phys. Lett.*, 81(2):289–291, 2002.

[34] K. Sethuraman, S. Ochiai, K. Kojima, and T. Mizutani. Performance of poly(3-hexylthiophene) organic field-effect transistors on cross-linked poly(4-vinyl phenol) dielectric layer and solvent effects. *Appl. Phys. Lett.*, 92(18):183302, 2008.

[35] M. Hwang, H. Lee, Y. Jang, J.H. Cho, S. Lee, D.H. Kim, and K. Cho. Effect of curing conditions of a poly(4-vinylphenol)gate dielectric on the performance of a pentacene-based thin film transistor. *Macromol. Res.*, 17:436–440, 2009.

[36] H.S. Byun, Y.X. Xu, and C.K. Song. Fabrication of high performance pentacene thin film transistors using poly(4-vinylphenol) as the gate insulator on

polyethyleneterephthalate substrates. *Thin Solid Films*, 493(12):278 – 281, 2005.

[37] Y. Choi, H. Kim, K. Sim, K. Park, C. Im, and S. Pyo. Flexible complementary inverter with low-temperature processable polymeric gate dielectric on a plastic substrate. *Org. Electr.*, 10(7):1209 – 1216, 2009.

[38] M.E. Roberts, N. Queralto, S.C.B. Mannsfeld, B.N. Reinecke, W. Knoll, and Z. Bao. Cross-linked polymer gate dielectric films for low-voltage organic transistors. *Chem. Mat.*, 21(11):2292–2299, 2009.

[39] W. Xu and S.W. Rhee. Hysteresis-free organic field-effect transistors with high dielectric strength cross-linked polyacrylate copolymer as a gate insulator. *Org. Elect.*, 11(5):836 – 845, 2010.

[40] W. Xu and S.W. Rhee. Low-operating voltage organic field-effect transistors with high-κ cross-linked cyanoethylated pullulan polymer gate dielectrics. *J. Mater. Chem.*, 19:5250–5257, 2009.

[41] W. Xu and S.W. Rhee. Organic field-effect transistors with cross-linked high-κ cyanoethylated pullulan polymer as a gate insulator. *Org. Electr.*, 11(6):996 – 1004, 2010.

[42] L.L. Chua, P.K.H. Ho, H. Sirringhaus, and R.H. Friend. High-stability ultrathin spin-on benzocyclobutene gate dielectric for polymer field-effect transistors. *Appl. Phys. Lett.*, 84(17):3400–3402, 2004.

[43] T.W. Lee, J.H. Shin, I.N. Kang, and S.Y. Lee. Photocurable organic gate insulator for the fabrication of high-field effect mobility organic transistors by low temperature and solution processing. *Adv. Mater.*, 19(18):2702–2706, 2007.

[44] F.Y. Yang, K.J. Chang, M.Y. Hsu, and C.C. Liu. High-performance poly(3-hexylthiophene) transistors with thermally cured and photo-cured PVP gate dielectrics. *J. Mater. Chem.*, 18:5927–5932, 2008.

[45] S.H. Lee, D.J. Choo, S.H. Han, J.H. Kim, Y.R. Son, and J. Jang. High performance organic thin-film transistors with photopatterned gate dielectric. *Appl. Phys. Lett.*, 90(3):033502, 2007.

[46] P.L. Egerton, E. Pitts, and A. Reiser. Photocycloaddition in solid poly(vinyl cinnamate). the photoreactive polymer matrix as an ensemble of chromophore sites. *Macrom.*, 14(1):95–100, 1981.

[47] J. Jang, S.H. Kim, S. Nam, D.S. Chung, C. Yang, W.M. Yun, C.E. Park, and J.B. Koo. Hysteresis-free organic field-effect transistors and inverters using photocrosslinkable poly(vinyl cinnamate) as a gate dielectric. *Appl. Phys. Lett.*, 92(14):143306, 2008.

[48] J. Jang, S. Nam, J. Hwang, J.J. Park, J. Im, C.E. Park, and J.M. Kim. Photocurable polymer gate dielectrics for cylindrical organic field-effect transistors with high bending stability. *J. Mater. Chem.*, 22:1054–1060, 2012.

[49] C.M. Wu, S.H. Su, H.T. Wang, M. Yokoyama, and S.L. Fu. Ultraviolet-patternable polymer insulator for organic thin-film transistors on flexible substrates. *Jap. J. Appl. Phys.*, 50(4):04DK23, 2011.

[50] S. Pyo, H. Son, K.Y. Choi, M.H. Yi, and S.K. Hong. Low-temperature processable inherently photosensitive polyimide as a gate insulator for organic thin-film transistors. *Appl. Phys. Lett.*, 86(13):133508, 2005.

[51] K.S. Jang, H.J. Suk, W.S. Kim, T. Ahn, J.W. Ka, J. Kim, and M.H. Yi. Direct photo-patternable, low-temperature processable polyimide gate insulator for pentacene thin-film transistors. *Org. Electron.*, 13(9):1665 – 1670, 2012.

[52] S.H. Kim, K. Hong, M. Jang, J. Jang, J.E. Anthony, H. Yang, and C.E. Park. Photo-curable polymer blend dielectrics for advancing organic field-effect transistor applications. *Adv. Mater.*, 22(43):4809–4813, 2010.

[53] T. Cahyadi, H.S. Tan, E.B. Namdas, S.G. Mhaisalkar, P.S. Lee, Z.K. Chen, C.M. Ng, and F.Y.C. Boey. Improved pentacene device characteristics with solgel SiO_2 dielectric films. *Org. Electron.*, 8(4):455 – 459, 2007.

[54] J. Tardy, M. Erouel, A.L. Deman, A. Gagnaire, V. Teodorescu, M.G. Blanchin, B. Canut, A. Barau, and M. Zaharescu. Organic thin film transistors with HfO_2 high-κ gate dielectric grown by anodic oxidation or deposited by solgel. *Microelectron. Reliab.*, 47(23):372 – 377, 2007.

[55] J. Ramajothi, S. Ochiai, K. Kojima, and T. Mizutani. Performance of organic field-effect transistor based on poly(3-hexylthiophene) as a semiconductor and

titanium dioxide gate dielectrics by the solution process. *Jap. J. Appl. Phys.*, 47(11):8279–8283, 2008.

[56] Z. Bao, V. Kuck, J.A. Rogers, and M.A. Paczkowski. Silsesquioxane resins as high-performance solution processible dielectric materials for organic transistor applications. *Adv. Funct. Mater.*, 12(8):526–531, 2002.

[57] S. Jeong, D. Kim, B.K. Park, S. Lee, and J. Moon. Influence of silanol groups on the electrical performance of organic thin-film transistors utilizing organosiloxane-based organic-inorganic hybrid dielectrics. *Nanotechnology*, 18(2):025204, 2006.

[58] B.H. Hamadani, D.A. Corley, J.W. Ciszek, J.M. Tour, and D. Natelson. Controlling charge injection in organic field-effect transistors using self-assembled monolayers. *Nano Lett.*, 6(6):1303–1306, 2006.

[59] M. Surin, P. Leclère, R. Lazzaroni, J.D. Yuen, G. Wang, D. Moses, A.J. Heeger, S. Cho, and K. Lee. Relationship between the microscopic morphology and the charge transport properties in poly(3-hexylthiophene) field-effect transistors. *J. Appl. Phys.*, 100(3), 2006.

[60] W.Y. Chou, C.W. Kuo, H.L. Cheng, Y.R. Chen, F.C. Tang, F.Y. Yang, D.Y. Shu, and C.C. Liao. Effect of surface free energy in gate dielectric in pentacene thin-film transistors. *Appl. Phys. Lett.*, 89(11), 2006.

[61] A. Salleo, M.L. Chabinyc, M.S. Yang, and R.A. Street. Polymer thin-film transistors with chemically modified dielectric interfaces. *Appl. Phys. Lett.*, 81(23):4383–4385, 2002.

[62] L.A. Majewski, R. Schroeder, M. Grell, P.A. Glarvey, and M.L. Turner. High capacitance organic field-effect transistors with modified gate insulator surface. *J. Appl. Phys.*, 96(10):5781–5787, 2004.

[63] D. Knipp, R.A. Street, A. Völkel, and J. Ho. Pentacene thin film transistors on inorganic dielectrics: Morphology, structural properties, and electronic transport. *J. Appl. Phys.*, 93(1):347–355, 2003.

[64] T.W. Kelley, L.D. Boardman, T.D. Dunbar, D.V. Muyres, M.J. Pellerite, and T.P. Smith. High-performance OTFTs using surface-modified alumina dielectrics. *J. Phys. Chem. B*, 107(24):5877–5881, 2003.

[65] T.C. Gorjanc, I. Lévesque, and M. D'Iorio. Organic field effect transistors based on modified oligo-p-phenylevinylenes. *Appl. Phys. Lett.*, 84(6):930–932, 2004.

[66] J. Veres, S.D. Ogier, S.W. Leeming, D.C. Cupertino, and S.M. Khaffaf. Low-κ insulators as the choice of dielectrics in organic field-effect transistors. *Adv. Funct. Mat.*, 13(3):199–204, 2003.

[67] L.L. Chua, J. Zaumseil, J.F. Chang, E.C. Ou, P.K. Ho, H. Sirringhaus, and R.H. Friend. General observation of n-type field-effect behaviour in organic semiconductors. *Nature*, 434(7030):194–199, 2005.

[68] J.W. Chang, W.L. Hsu, C.Y. Wu, T.F. Guo, and T.C. Wen. The polymer gate dielectrics and source-drain electrodes on n-type pentacene-based organic field-effect transistors. *Org. Elect.*, 11(10):1613 – 1619, 2010.

[69] S. Lee, B. Koo, J. Shin, E. Lee, H. Park, and H. Kim. Effects of hydroxyl groups in polymeric dielectrics on organic transistor performance. *Appl. Phys. Lett.*, 88(16), 2006.

[70] S. Kobayashi, T. Nishikawa, T. Takenobu, S. Mori, T. Shimoda, T. Mitani, H. Shimotani, N. Yoshimoto, S. Ogawa, and Y. Iwasa. Control of carrier density by self-assembled monolayers in organic field-effect transistors. *Nature Mat.*, 3(5):317–322, 2004.

[71] K.P. Pernstich, S. Haas, D. Oberhoff, C. Goldmann, D.J. Gundlach, B. Batlogg, A.N. Rashid, and G. Schitter. Threshold voltage shift in organic field effect transistors by dipole monolayers on the gate insulator. *J. Appl. Phy.*, 96(11):6431–6438, 2004.

[72] A. Stadler and M. Burghart. Organic-MISTFT capacitance measurements. *IEEE T. Dielect. El. In.*, 13(4):850–854, 2006.

[73] Ni. Zhao, O. Marinov, G.A. Botton, M.J. Deen, B.S. Ong, Y. Wu, and P. Liu. Characterization of MOS structures based on poly (3,3"'-dialkyl-quaterthiophene). *IEEE T. Electron. Dev.*, 52(10):2150–2156, 2005.

[74] Y. Taur and T.H. Ning. *Fundamentals of Modern VLSI Devices*. Cambridge University Press, 2009.

[75] L.M. Terman. An investigation of surface states at a silicon/silicon oxide interface employing metal-oxide-silicon diodes. *Solid State Electron.*, 5(5):285–299, 1962.

[76] J.E. Puskas, T.E. Long, and R.F. torey. *In Situ Spectroscopy of Monomer and Polymer Synthesis.* Kluwer Academic/Plenum Publishers, 2003.

[77] D. Hayward, A.J. Mackinnon, S. Radhakrishnan, and R.A.Pethrick. Investigation of cure in epoxy acrylate resins using rheological and dielectric measurements. *Polym. Eng. Sci.*, 35(2):184–189, 1995.

[78] T.Y. Lee, C.A. Guymon, E.S. Jönsson, and C.E. Hoyle. The effect of monomer structure on oxygen inhibition of (meth)acrylates photopolymerization. *Polymer*, 45(18):6155–6162, 2004.

[79] P. Glöckner. *Radiation curing.* William Andrew Publishing, 2008.

[80] K. Suyama and M. Shirai. Photobase generators: Recent progress and application trend in polymer systems. *Prog. Polym. Sci.*, 34(2):194–209, 2009.

[81] J.V. Crivello and J.H.W. Lam. Diaryliodonium salts: A new class of photoinitiators for cationic polymerization. *Macromol.*, 10(6):1307–1315, 1977.

[82] J.V. Crivello. The discovery and development of onium salt cationic photoinitiators. *J. Polym. Sci. A Polym. Chem.*, 37(23):4241–4254, 1999.

[83] P. Glöckner, T. Jung, and S. Struck. *Radiation Curing: Coatings and Printing Inks, Technical Basics, Applications and Trouble Shooting.* Vincentz Network, 2009.

[84] C. Decker and K. Moussa. Kinetic study of the cationic photopolymerization of epoxy monomers. *J. Polym. Sci. A Polym. Chem.*, 28(12):3429–3443, 1990.

[85] J.P. Fouassier and J. F. Rabek. *Radiation curing in polymer science and technology - volume IV: Practical aspects and applications.* Elsevier Science publisher, Barking, 1993.

[86] J.A. Brydson. *Plastics materials.* Butterworth Heinemann, Oxford, 1999.

[87] C.E. Barnes. Method of polymerizing unsaturated methylene compounds. *U.S. Patent 2369520*, 13 Feb 1945.

[88] L.M. Richards. Photopolymerization of vinylidene compounds in the presence of n-di-substituted dithiocarbamic esters. *U.S. Patent 2423520*, 8 Jul 1947.

[89] C.L. Agre. Process of photopolymerization. *U.S. Patent 2367661*, 23 Jan 1945.

[90] J.P. Fouassier and J.F. Rabek. *Radiation curing in polymer science and technology - volume I: Fundamentals and methods.* Elsevier Science publisher, Barking, 1993.

[91] J. Zhang, C.M. Li, M.B. Chan-Park, Q. Zhou, Y. Gan, F. Qin, B. Ong, and T. Chen. Fabrication of thin-film organic transistor on flexible substrate via ultraviolet transfer embossing. *Appl. Phys. Lett.*, 90(24):243502–243502, 2007.

[92] C. Belon, X. Allonas, C. Croutx-barghorn, and J. Laleve. Overcoming the oxygen inhibition in the photopolymerization of acrylates: A study of the beneficial effect of triphenylphosphine. *J. Polym. Sci. A Polym. Chem.*, 48(11):2462–2469, 2010.

[93] M.A. Gauthier, I. Stangel, T.H. Ellis, and X.X. Zhu. Oxygen inhibition in dental resins. *J. Dent. Res.*, 84(8):725–729, 2005.

[94] J.P. Fouassier and J.F. Rabek. *Radiation curing in polymer science and technology - volume III: Polymerization mechanisms.* Elsevier Science publisher, Barking, 1993.

[95] M.S. Kharasch, E.M. May, and F.R. Mayo. The addition of thiocyanic acid to olefinic double bonds. *J. Am. Chem. Soc.*, 59(8):1580–1580, 1937.

[96] C.S. Marvel and R.R. Chambers. Polyalkylene sulfides from diolefins and dimercaptans. *J. Am. Chem. Soc.*, 70(3):993–998, 1948.

[97] C.E. Hoyle, T.Y. Lee, and T. Roper. Thiol-enes: Chemistry of the past with promise for the future. *J. Polym. Sci. A Polym. Chem.*, 42:5301–5338, 2004.

[98] A.K. O'Brien, N.B. Cramer, and C.N. Bowman. Oxygen inhibition in thiol-acrylate photopolymerizations. *J. Polym. Sci. A Polym. Chem.*, 44(6):2007–2014, 2006.

[99] C.R. Morgan, F. Magnotta, and A.D. Ketley. Thiol/ene photocurable polymers. *J. Polym. Sci. Polym. Chem. Ed.*, 15(3):627–645, 1977.

[100] G. Socrates. *Infrared characteristic group frequencies.* John Wiley and Sons, Chichester, 1980.

[101] J. Endo, T. Matsumoto, and J. Kido. Organic electroluminescent devices with a vacuum-deposited lewis-acid-doped hole-injecting layer. *JPN J. Appl. Phys.*, 41(Part 2, No. 3B):L358–L360, 2002.

[102] H. Etschmaier, P. Pacher, A. Lex, G. Trimmel, C. Slugovc, and E. Zojer. Continuous tuning of the threshold voltage of organic thin-film transistors by a chemically reactive interfacial layer. *Appl. Phys. A Mat. Sci. Proc.*, 95:43–48, 2009.

[103] J.Liu, L.N. Lewis, and A.R. Duggal. Photoactivated and patternable charge transport materials and their use in organic light-emitting devices. *Appl. Phys. Lett.*, 90(23):233503, 2007.

[104] Z. Fahem and W. Bauhofer. Free radical fast photo-cured gate dielectric for top-gate polymer field effect transistors. *Organic Electronics*, 13(8):1382 – 1385, 2012.

[105] K. Studer, C. Decker, E. Beck, and R. Schwalm. Overcoming oxygen inhibition in UV-curing of acrylate coatings by carbon dioxide inerting, part I. *Prog. Org. Coat.*, 48(1):92–100, 2003.

[106] D. Wang, L. Carrera, and M.J.M. Abadie. Photopolymerization of glycidyl acrylate and glycidyl methacrylate investigated by differential photocalorimetry and FT-IR. *Eur. polym. J.*, 29(10):1379–1386, 1993.

[107] A. Wang, I. Kymissis, V. Bulovic, and A. I. Akinwande. Tunable threshold voltage and flatband voltage in pentacene field effect transistors. *Appl. Phys. Lett.*, 89:113109, 2006.

[108] J.M. Choi. Correlation between the interface states and ultraviolet treatment for low-voltage pentacene thin-film transistors. *Syn. Met.*, 159:1689–1693, 2009.

[109] V. Starov, N. Ivanova, and R.G. Rubio. Why do aqueous surfactant solutions spread over hydrophobic substrates? *Adv. Coll. Inter. Sci.*, pages 153–162, 2010.

[110] S. Scheinert and G. Paasch. Interdependence of contact properties and field- and density-dependent mobility in organic field-effect transistors. *J. Appl. Phys.*, 105(1):014509–014509, 2009.

[111] Z. Fahem and W. Bauhofer. Thiol-ene polymer based fast photo-curable gate insulator for organic field effect transistors. *Microelectron. Eng.*, 105:74–76, 2013.

[112] U. Müllera, A. Kunzea, C. Herzigb, and J. Weisb. Photocrosslinking of silicones. Part 13. Photoinduced thiol-ene crosslinking of modified silicones. *J.M.S. Pure Appl. Chem. A*, 33:439–457, 1996.

[113] J.S. Kim, S. Yang, H. Park, and B.S. Bae. Photo-curable siloxane hybrid material fabricated by a thiolene reaction of solgel synthesized oligosiloxanes. *Chem. Commun.*, 47:6051–6053, 2011.

[114] N.B. Cramer and C.N. Bowman. Kinetics of thiol-ene and thiol-acrylate photopolymerizations with real-time fourier transform infrared. *J. Polym. Sci. A Polym. Chem.*, 39:3311–3319, 2001.

[115] R. Schwalm. *UV Coatings: Basics, Recent Developments and New Applications.* Elsevier Science, 2006.

[116] M. Egginger, S. Bauer, R. Schwödiauer, H. Neugebauer, and N.S. Sariciftci. Current versus gate voltage hysteresis in organic field effect transistors. *Chemical Monthly*, 140(7):735–750, 2009.

[117] M. Egginger, M. Irimia-Vladu, R. Schwödiauer, A. Tanda, I. Frischauf, S. Bauer, and N.S. Sariciftci. Mobile ionic impurities in poly(vinyl alcohol) gate dielectric: Possible source of the hysteresis in organic field-effect transistors. *Adv. Mater.*, 20(5):1018–1022, 2008.

Acknowledgements

This work was achieved at the Institute of Optical and Electronic Materials at the Hamburg University of Technology. I would like to express my sincerest thanks to many people who made this work possible.

First, I am grateful to my advisor, Prof. Dr. Wolfgang Bauhofer, for his advices and comments. I appreciate the freedom he gave to me during my work. I would like also to thank Prof. Dr.-Ing. Wolfgang Krautschneider and Prof. Dr. Manfred Eich for their contributions to this study.

I thank Dr. Altan Yildirim for introducing me in the field of organic electronics. I am grateful to Ricardo Starbird for introducing me in the spectroscopic techniques. I would like to thank my colleagues at the Institute of Optical and Electronic Materials: Cenk Akbulak, Nils Steller, Stefan Prorok, Michel Castellanos, Dr. Ali Eken, Dr. Alexander Petrov, Dr. Wenjing Li, Jan Hampe, Sebastian Jakobs, Dirk Jalas, Hooi Sing Lee and Dr. Jan Hendrik Wülbern for the pleasant and fruitful work atmosphere. A special thank goes to Gabriele Birjukov for her help in administrative issues even outside the university. I thank Iris Bucher, Stefan Schön and Michael Seiler for their technical support. I thank also Volkmar Block, Sabine Kölling and Joachim Kunze from the central laboratory of the TUHH for the explanation of the different analytical tools and for the characterization of PETMP.

Finally, my deepest gratitude goes to my family, especially to my son and wife for their support, patience and understanding. I thank my parents for their advices and encouragement.

List of publication

Journal articles:

Z. Fahem and W. Bauhofer. Free radical fast photo-cured gate dielectric for top-gate polymer field-effect transistors. Organic Electronics, 13(8):1382-1385, 2012.

Z. Fahem and W. Bauhofer. Thiol-ene polymer based fast photo-curable gate insulator for organic field-effect transistors. Microelectronic Engineering, 105:74-76, 2013.

Conferences:

Z. Fahem and W. Bauhofer. Bias stress measurement in organic MIS-capacitors. International Conference of Organic Electronics 2010, Paris, France.

Z. Fahem, N. Steller and W. Bauhofer. Low-temperature curable organic-inorganic hybrid dielectric for organic field-effect transistors. International Conference of Organic Electronics 2012, Tarragona, Spain.

Z. Fahem and W. Bauhofer. Fast photo-cured gate dielectric for organic field-effect transistors. International Conference of Organic Electronics 2012, Tarragona, Spain.

Curriculum Vitae

Name	Zied Fahem
Date of birth	18.09.1981
Place of Birth	Tunis (Tunisia)
Nationality	Tunisian

Education

2000	High-school diploma at Lycée Pilote Gafsa
2000-2001	German course at the Studienkolleg at Hanover University
2001-2003	Vordiplom of Electrical and communication engineering at Hanover University
2003-2007	Dipl.-Ing. of Electrical and communication engineering at TU Munich
2008-2013	Ph.D. student in Electrical and communication engineering at TU Hamburg-Harburg

Professional experience

2007	Application engineer at Nanotec GmbH
2007-2008	Research assistant at TU Chemnitz
2008-2010	Research assistant at TuTech Innovation GmbH, Hamburg
2010-2012	Research assistant at TU Hamburg-Harburg